THE
FISHERIES OF EUROPE

BELL'S ADVANCED ECONOMIC GEOGRAPHIES

General Editor

PROFESSOR R. O. BUCHANAN
M.A.(N.Z.), B.Sc.(Econ.), Ph.D.(London)
Professor Emeritus, University of London

A. *Systematic Studies*

AN ECONOMIC GEOGRAPHY OF OIL
Peter R. Odell, B.A., Ph.D.

PLANTATION AGRICULTURE
P. P. Courtenay, B.A., Ph.D.

NEW ENGLAND: A STUDY IN INDUSTRIAL ADJUSTMENT
R. C. Estall, B.Sc.(Econ.), Ph.D.

GREATER LONDON: AN INDUSTRIAL GEOGRAPHY
J. E. Martin, B.Sc.(Econ.), Ph.D.

GEOGRAPHY AND ECONOMICS
Michael Chisholm, M.A.

AGRICULTURAL GEOGRAPHY
Leslie Symons, B.Sc.(Econ.), Ph.D.

REGIONAL ANALYSIS AND ECONOMIC GEOGRAPHY
John N. H. Britton, M.A., Ph.D.

THE FISHERIES OF EUROPE: AN ECONOMIC GEOGRAPHY
James R. Coull, M.A., Ph.D.

A GEOGRAPHY OF TRADE AND DEVELOPMENT IN MALAYA
P. P. Courtenay, B.A., Ph.D.

B. *Regional Studies*

AN ECONOMIC GEOGRAPHY OF EAST AFRICA
A. M. O'Connor, B.A., Ph.D.

AN ECONOMIC GEOGRAPHY OF WEST AFRICA
H. P. White, M.A., & M. B. Gleave, M.A.

YUGOSLAVIA: PATTERNS OF ECONOMIC ACTIVITY
F. E. Ian Hamilton, B.Sc.(Econ.), Ph.D.

RUSSIAN AGRICULTURE: A GEOGRAPHIC SURVEY
Leslie Symons, B.Sc.(Econ.) Ph.D.

AN AGRICULTURAL GEOGRAPHY OF GREAT BRITAIN
J. T. Coppock, M.A., Ph.D.

AN HISTORICAL INTRODUCTION TO THE ECONOMIC GEOGRAPHY
OF GREAT BRITAIN
Wilfred Smith, M.A.

THE BRITISH IRON & STEEL SHEET INDUSTRY SINCE 1840
Kenneth Warren, M.A., Ph.D.

THE
FISHERIES OF EUROPE

AN ECONOMIC GEOGRAPHY

JAMES R. COULL
M.A., Ph.D.

Senior Lecturer in Geography
University of Aberdeen

LONDON
G. BELL & SONS, LTD.
1972

ISBN 0 7135 1612 7

Printed in Great Britain by
NEILL AND CO. LTD., EDINBURGH

TO
MY PARENTS

Contents

Maps and Diagrams

ix

Preface

In the preparation of this book, my thanks are due to a large number of people in the academic world, in the fishing industry, and in government departments. I am especially grateful in the first place to Professor R. O. Buchanan, whose constant encouragement and advice have played no small part in the completion of the work. Professor K. Walton, Head of the Geography Department here in the University of Aberdeen, has also been a great source of encouragement with his interest in maritime affairs; and several of the scientific officers of the Torry Marine Laboratory in Aberdeen have given invaluable advice and information on matters of marine biology.

The drawing of the maps and diagrams was kindly undertaken by Mr. M. Wood, Mr. C. Bremner, Miss A. Watt and Miss D. Low, and I am indebted to Mrs. A. Reid, Miss Y. Wilson and Mrs. M. Riddoch for preparing the typescript. For the obtaining of source of material I have to thank the staff of the University Library in Aberdeen, and also Mr. H. McCall and his staff of the library of Torry Marine Laboratory. Mr. R. Aitken has kindly undertaken the work of proof-reading.

It scarcely needs stating that the undertaking of any work on a continental scale has essential limitations; but at the same time a limited amount has been written about fisheries, and this book sets out to explore the relationships within the fishing industry of Europe from the geographical viewpoint. While I have sought to develop a systematic—rather than a regional—approach, the considerable variations in fisheries within Europe have also necessitated treatment. The space devoted to the fisheries of different regions and countries is not, however, proportional to their economic importance. The main reason for this is the unequal amount of published data; but the frequency of examples from Scotland and Norway is also related to my own greater familiarity with the situations in these two countries.

I am indebted to The John Hopkins Press for their kind permission to adapt Figures 1 and 2 from diagrams in their book

The Common Wealth in Ocean Fisheries by F. Y. Christy and A. Scott, which was printed in the series *Resources for the Future, Inc.* I am also indebted to Fishing News Books Ltd., for permission to adapt biomass data from *Natural Bases of Fisheries in the Atlantic Ocean* by T. Laevastu, in the book *Atlantic Ocean Fisheries*, edited by G. Borgstrom and A. J. Heighway.

Finally, it has been a privilege for one whose origins are among the fisher folk to set something of the special characteristics and problems of the fisheries before a wider public.

Department of Geography   James R. Coull
University of Aberdeen
April, 1971

CHAPTER I

Introduction

The fisheries prosecuted by the European nations constitute an important part of the world's effort in the industry. The total catch of the continent (including the European part of the U.S.S.R.) now exceeds 15,000,000 metric tons per year, and is about one-quarter of the world total. For lack of complete data, it is not possible to state the value of the European part of the catch, but it is certainly in excess of one-quarter of the total value. While on the global plane Peru now leads all other countries in tonnage landed, and Japan is an easy first in value of catch, Europe has six of the leading fifteen fishing nations, which are those which land over c. 800,000 metric tons per year; and this does not include the European part of the U.S.S.R., at whose ports landings are more than double this figure.

In the major fishing areas of the globe for which the Food and Agricultural Organisation of the United Nations (F.A.O.) collects statistics, the North-East Atlantic Area (whence comes over 60 per cent. of the European catch) competes for leadership in yield with the West Central Pacific (mainly exploited by Japan) and the South-East Pacific (dominated by Peru). Although physical yield is now intermediate between those of these other areas, the value of the catch from the North-East Atlantic is greater than that of either.

It should be stated, however, that, although in the front rank in sea fisheries, Europe is of very limited importance in the yield of inland waters (rivers and lakes), whence comes a significant proportion (c. 12 per cent.) of the world total; of this total Europe contributes only c. 1 per cent.

The European continent presents a great variation in opportunities in fisheries, and in the stage of their development. At one extreme there are the landlocked states like Switzerland and Hungary, which can exploit only their inland waters, while at the other are countries like Norway, Iceland, Portugal and the United Kingdom, with long coast lines fronting the open sea, which contains some of the world's richest fishing grounds. Intermediate in opportunity are such nations as Finland, Italy and Bulgaria, whose only immediate access is to the generally poorer resources of the enclosed Baltic, Mediterranean and Black Seas.

Europe also shows differences in outlook upon fisheries development which echo the major contrasts of the world. The main concern of the advanced nations, of the north-west seaboard of Europe especially, has become that of the long-term conservation and management of a resource tending to become over-exploited, rather than the expansion of their catch. Of principal importance in southern and south-east Europe, on the other hand, is the employment of known techniques to increase output, food supplies and employment, and to organize the distribution of the catch. Further, the U.S.S.R., Poland and East Germany hardly fall into either of these groups: here the immense post-war expansion of fisheries has been promoted by the state authorities because the investment involved produces greater and more rapid returns of flesh foods than investment in farming. The modern expansion of Soviet fisheries especially has been far greater than that of any other European nation, and they now exceed those of any other European country.

While the absolute importance of the fisheries of Europe has been continually increasing over the modern period since the Industrial Revolution—and indeed more irregularly for centuries previously—the outstanding feature of the mid-twentieth century has been the rise of rivals in other continents and the decline in the relative importance of Europe; and this has occurred despite accelerating development within Europe, and despite the fact that many of the innovations introduced have originated among her nations. The early twentieth century witnessed the upsurge of Japan, the United States and Canada as fisheries nations employing improved methods, and these have since been joined by others, including the U.S.S.R., Peru and South Africa. It should also be noted that fisheries production has continued high among the other nations of Monsoon Asia, especially India and China; and in this region, the rivers and coastal waters of which were historically the main fishing area of the world, production is now rising, although productivity is still far behind that of the leading nations.

In Europe itself the reasons for the importance of the fisheries are basically two-fold. In the first place, the fish stocks in the seas off the western seaboard, and extending outwards into the north-west Atlantic and Arctic Oceans, constitute a resource base with few rivals on the globe; and fish constitute a significant element in the food supplies of millions of people in Europe. Secondly, there is the store of technical knowledge and commercial experience in fisheries in the continent, built up especially during the last century and a half, but with origins going back to medieval times. Thus geographical and historical factors have blended and inter-acted in the rise of the fisheries, and in their continued development and exploitation.

THE APPROACH OF THE ECONOMIC GEOGRAPHER TO THE STUDY OF FISHERIES

The geographer studying fisheries is essentially interested in the manner in which they are influenced or conditioned by environmental factors, and in elucidating the distribution patterns of phenomena and activity developed in their exploitation. He will have to assess the character of both sea and land, and his field will be one involving principally the convergence of marine ecology and economic geography. The economic geographer will view the environmental factors with a view to measuring their effect in terms of cost; and he will proceed to analyze the extent to which distribution of fishing effort, of shore facilities and of final market destinations are determined by costing advantages. In so doing, it will become apparent that the biotic resource itself is subject to fluctuations, and this variable interacts with economic variables such as, for example, those embodied in the trade cycle. Fish also include a wide range of species, with big variations between them in the scale and the pattern of supply and demand. The differences in value between species can be several thousand per cent., and even for a single species the value can vary several hundred per cent. Also basic is the fact that fish are largely a common property international resource of the high seas, exploited by techniques of hunting rather than of husbanding, and are highly perishable. These are weighty factors in the economics of the industry, and when superimposed on the biological fluctuations they tend to militate against stable patterns of utilization and trade, and against economic efficiency.

As in so many kinds of study—in geography above all—there are great virtues in a regional analysis of fisheries, and Europe here is in considerable measure self-contained. The north-east Atlantic is exploited almost entirely by European nations, although towards the north-west part of the ocean European fleets share the fisheries with those of Canada and the U.S.A., while in the south (including the Mediterranean) there is an overlap in effort with the countries of North Africa. The extension of fishing effort into the Southern Atlantic, especially by the U.S.S.R., has meant that the European sphere in fisheries now overlaps those of the nations of West and Southern Africa, of the Latin American countries and of Japan. Until recent years, trade in fish and fish products has for the great part circulated within Europe, although now a not insignificant part of the scene is the import of fish meal from Peru. For detailed discussion of fisheries and trade within Europe, concentration must be on the national units, as each country has its own systems of costs and incentives, and its own circumstances of wages and prices.

STATISTICS & STRUCTURE OF THE INDUSTRY

Any attempt to formulate the economic geography of fisheries is inevitably hampered by the limitations of available data and statistics. While it must be rare, if not unknown, for the economic geographer to find data sufficiently detailed, and appropriately organized, to suit his purpose fully in any field, fisheries present an unusually difficult problem, although, thanks to the work of F.A.O. and other international organizations the situation is certainly improving. While nearly all the European nations publish reports and records of their fisheries, their systems are each adapted to national needs, and detailed comparison is seldom possible. There is the further complication that the U.S.S.R. has considerable fisheries in the Asiatic section of the country which are included within national statistics. For individual countries data on landings at the ports are often available over a considerable period, but less is known of where the fish are caught and of how they are disposed of after landing. It is especially difficult to measure adequately the effort put into fishing, and to analyze the economic relationships involved in the necessary ancillary shore services and marketing chains.

Whatever the difficulties in econometrics in fisheries, the essential structure of the industry is at least clear in qualitative terms. In the first place there is the catching sector of the industry, which involves vessels going out from their operating bases to fishing grounds and returning with their catches. Secondly, there is the sector engaged in marketing fish in the ports. The fish may then be processed; finally, fresh or processed, it is distributed to consuming points. The basic resource consists of the fish stocks, and capital and labour are employed in varying proportions in the fishing fleets and in the ancillary shore installations and marketing chains. There are naturally subsidiary industries and occupations related to the fisheries, concerned with such activities as providing boats, gear and equipment, processing plant and haulage.

FISH STOCKS & FISHING OPERATIONS

The resource base of the fish stocks is a variable in both place and time as well as in value. The existence of the continental shelves around the North Atlantic does in large measure define the location of exploitable fish stocks, although pelagic species especially may be found in deeper waters. On the continental shelves themselves, however, there are marked variations in abundance and in composition of stocks on different grounds; and these variations are related both to the character of the marine environment and to the intensity of man's exploitation of it. A large part of the variations—especially of pelagic fish— is seasonal; and there are also significant year-to-year changes in abundance. To understand these periodic and spatial variations the geographer must have recourse to the findings of the science of marine biology. The availability of the resource is conditioned by the abundance of the vegetation of the sea—the phytoplankton, which is in turn related to the factors of physical

B

oceanography, especially those which determine the availability of nutrients. The scale of the resource is also determined by the structure of the food chains and food pyramids of the sea; by the reproduction and growth rates of different fish species; by fish migrations, which are largely seasonal spawning movements; and, in modern times, by man's rate of exploitation, to name only the main factors. In short, the sea—as the environment in which the fish occur—must be viewed as an ecosystem. For the economic geographer, the study of these phenomena would lead ideally to a continuous year-to-year record of the fish stocks, together with the defining of optimum rates of exploitation; in practice, however, it is only in exceptional cases that this can be approached.

The operation of the fishing fleets will be a subject for study with a view to assessing their profitability and economic efficiency. These will be determined in part by factors of physical geography—by the incidence of gales and poor visibility inhibiting fishing, by floating ice on the Arctic margins and on occasion by complications caused by tides. From the assessment of profitability of fleets, the ideal would be to proceed to compare the yield from different fish stocks and grounds, and of different fishing gears and methods: yield per unit of invested capital and per unit of labour will be valuable comparative criteria. The units involved in harvesting the sea are generally individual—and independent—fishing craft, although a proportion are organized in different degrees into fleets. The type of organization depends principally on the distance from base at which grounds are worked: local inshore grounds are usually fished by smaller independent craft, while the larger vessels which exploit distant waters are generally owned by companies with a unified direction of operations; and the big Soviet fleets may be complete with factory-ships and other supporting vessels, with management the responsibility of a government department.

The fact that a large proportion of the enterprise is still in the hands of independent small operators makes it difficult to get an all-over impression of European fisheries in any detail. The costing of operations is generally known for company-owned fleets, as records are kept which allow the assessment of such items as working capital, labour input, and profitability, but there may be no real records for small operators, and what figures there are often make estimates of profitability hazardous. While the situation is tending to improve as the smaller operators now often have aid from state funds, the differences in return as between boats and between years are probably considerably greater than for large craft, and accurate averages are difficult to compute. Another difficulty with the small operators is the assessment of the strength of the labour force: a large part of this is part-time, and included within this term can be those who perhaps stop fishing for three or four weeks per year to attend to a small holding, while at the other extreme it can involve men for whom fishing is a spare-time ancillary to other full-time employment. Although frequently short of capital, small inshore operators have one great advantage over the distant-water fleets in proximity to their grounds. The bases of these fleets may be over 1,000 miles from the grounds exploited, and the vessels often spend one-third or more of their time at sea in making the return passage.

PORTS & THEIR HINTERLANDS

In the investigation of the location and the distribution of activity in the shore sector of the fishing industry, it is well to begin at the operating bases, which can vary from small coastal settlements with no harbour, to the specialized docks and other provision made for fishing vessels at such ports as Hull, Bremerhaven, Esbjerg, Boulogne and Murmansk. It will be necessary to examine how far the location and the scale of activity at these bases are determined by costing advantages. Their distribution around the coasts may be plotted, and their spatial relationships with both fishing grounds and inland markets examined. The activities at the ports are dominated by their fish markets, although processing is also frequently of major importance; but also to be found is a complex of ancillary activities and establishments. These include marine engineering, the manufacture of fishing gear, the provision of specialized equipment like engines, winches and fishing and navigation aids. There may also be fishing vessel construction, while ship chandlers dealing with the specialized needs of the fleets are indispensable. Included also are the administrative offices which participate in organising and directing the effort of fishing and catering for the fleets and markets. From the scale and range of these various activities it will be possible to formulate a hierarchy of fishing ports.

The links of the ports with their hinterlands are generally most clearly shown by the marketing chains through which fish and fish products reach their points of consumption; but they are also shown in the reciprocal trade, in which foodstuffs, specialized equipment for the fleets and processing plants, and other items reach the ports. The trade patterns in fish in Europe are for the most part characteristic of those of economically advanced countries in being highly complex and it is only in part that the concept of ports and hinterlands is applicable. In the first place fish includes a great variety of species of different value, which may be marketed fresh and whole, or have value added by processes of filleting, freezing, smoking, curing, or of reduction to meal and oil; and fish is commonly sent from one port to another for marketing or processing. Secondly, particular gear and equipment types manufactured at individual centres may have a national—or international—distribution. Thirdly, in following seasonal fisheries, or in response to other variations, fleets may operate from different parts of the coasts at different times. Fourthly, business enterprise in fishing concerns may be linked or integrated with other activities unrelated to fisheries, and hence management decisions may partly depend on considerations unconnected with the industry. Finally, the location of operations and the direction of trade flows may be influenced by national policies and decisions, implemented by government departments. A main aim of the economic geographer will be to show and explain trade flows in fish and fish products. These will be related to differences in total and per capita consumption as between countries, districts, and towns; and the whole will show a pattern of regional variation in supply and demand.

Although the distinguishing of patterns of economic activity in their environmental context at the present time is essential, it must be realized that these are themselves dynamic. Change indeed permeates the industry, and is a function of the rising living standards and the developing technology and organization of the modern world. Hence awareness, and indeed analysis, of trends is necessary: developments in fisheries since the Second World War have been especially far-reaching, and have generated serious pressure on the basic resource, especially with the rise of big 'industrial' fisheries for reduction of catches to meal and oil.

FISHERIES IN THE GENERAL ECONOMIC CONTEXT

In addition to the study of fisheries in themselves, there is the task of assessing their role in wider economies, at regional, national and international levels. Here the character of the land bases will have an obvious importance, and fisheries will tend to be of greatest relative importance when a land base with few economic resources is in juxtaposition with rich offshore fishing grounds, as is the case with Iceland and Norway. A better endowed, wealthy hinterland, however, may stimulate fisheries by the market they provide, as occurs in Britain and France.

The economic importance of fisheries can be demonstrated by the assessment of their fraction in gross national products. It will be significant also to measure and compare the part played by fish in food supplies, and also to show how far consumption comes from the local or national catch, and the volume and value of any import or export balance.

In the relationship of fishing to other economic activities, of growing moment is the part played by the agencies of national governments, which is a reflection of the general increase in the public sector of national expenditure, and in the related growing role of planning in economic development. Fishing suffers with other industries of primary production in the general lag of price rises for its products behind those of manufacturing industry, with the resultant disproportionate increases in cost and reduction of profit margins. The role of government is usually initially seen as helping the industry to become solvent, but there is a tendency for state aid to become at least semi-permanent. In practice, the role of governments is most seen in various schemes to promote and stabilize fisheries by means of providing capital in the form of loans and grants, in guaranteeing loan repayments and fish prices, and in some operating subsidies. Some of these schemes are in themselves areally selective within countries, as for instance in stimulating development in Northern Norway and Southern Italy.

Various political pressures are inextricably bound up with this economic role played by the state; and political relationships are more obviously seen on the international scene. Fundamentally, the fisheries of Europe depend chiefly on an international biotic resource, which is already showing symptoms of over-exploitation : the operation of the Law of Diminishing Returns can be discerned. It is already apparent that the existing measures of regulations for mesh sizes of nets (to limit the catching of immature fish), and of extended national fishing limits are inadequate to meet the situation; in the long term the only solution is the limitation of fishing effort, although yields may be enhanced by extensive rearing of young fish to an age beyond the critical infant stages. The present situation presents a formidable challenge which can be met only by international agreements to resolve conflicts and promote conservation, and these must take into account the principles of both marine ecology and economics. Along with other physical and social scientists, the geographer has a role to play here, in clarifying environmental factors in operation on both sea and land, and in assessing factors which are areally significant.

CHAPTER II

The Principles of Fisheries Economics and their Application in Europe

An understanding of the economic geography of European—or any other—fisheries necessarily involves also an understanding of the basic principles of fisheries economics. Considerations of profitability will largely determine the intensity with which the resource is exploited, and the factors that govern profitability usually vary regionally. Fisheries economics is a relatively new academic study, largely because of the lack until recently of adequate recording for even basic analysis; and the special characteristics of fisheries as an industry, occasioned by the nature of the resource and the arrangements that have grown up to exploit it, make fisheries economics a specialist subject in its own right. The marketing and distribution sector of the industry shows economic relationships which are less distinctive, although it, too, has its own characteristics. Throughout the industry, there are generally marked short-term fluctuations in the tempo of activity; and variations in day-to-day and seasonal landings make themselves felt throughout the industry in proportionately high overhead costs.

Very generally, fisheries depend on a common property resource, and entry to them is in principle unrestricted. A fishery constitutes a self-sustaining 'fund' resource, the nature of which is modified by exploitation, and which can be drawn on so rapidly that it cannot fully replace itself. In the absence, however, of the restraints that would govern the amount of effort applied to an individually owned (or rented) resource, the tendency is for new producers to enter the industry as long

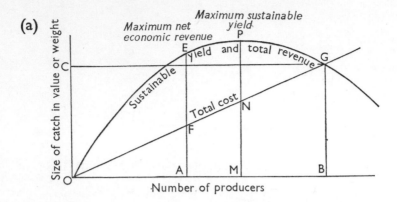

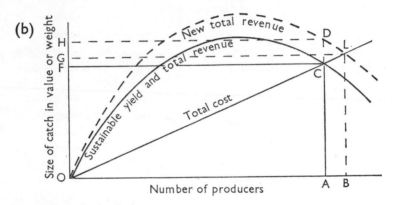

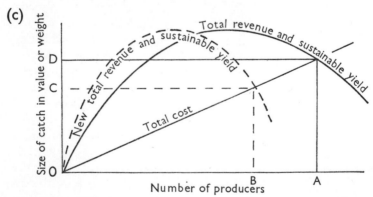

Figure 1. Curves of maximum sustainable yield:
(a) Total revenues, costs and sustainable yields with respect to effort.
(b) Effects of increased prices on yields and revenue curves.
(c) Effects of technological innovation on yield and revenue curves.

as any profit is to be made. The equilibrium situation is then reached only when the net yield (i.e. value of product less costs) is zero.[1] Hence the development of fishery in a situation of free competition tends to lead to an economically precarious or unviable situation. In a pioneer analysis, Professor H. Scott Gordon saw the roots of inherent instability of the fisheries in the fact that, as common property, they yielded no economic rent, which would have been a constraint on effort. On the other hand, it is only in inland waters that it is feasible—or economically possible—to expropriate the resource.

The manner in which a fish stock is influenced by exploitation involves its population dynamics, mainly through recruitment and mortality, and can be shown by considering the opening up of a new fishery.[2] Figure 1 shows the sequence of development as increasing numbers of producers enter the fishery; the sustainable yield and revenue at first rise steeply, but with progressive depletion of the stock through the increased mortality caused by fishing, a stage is ultimately reached where increased numbers of fishermen mean a decline in the total yield. The total cost of production, however, rises in proportion to the number of fishermen, and the equilibrium point is ultimately reached where the graph of cost intersects that of yield and the net yield is then zero. In Figure 1(a), it can be seen that a point of maximum sustainable yield (P) is reached at the highest point of the curve. At this point the greatest total weight of fish is produced, and the population structure of the stock is such that a maximum of the food is used for growth, as opposed to maintenance: in a virgin stock with a bigger proportion of older adult fish, the stock makes less efficient use of available food supplies as more is used for maintenance. Hence the biological optimum in a fishery occurs at the point of maximum sustainable yield, and this has tended to be the desirable goal in the eyes of marine biologists, and indeed, of national and international legislators. The maximum economic yield (denoted by E on Figure 1(a)), however, occurs where the distance between the graphs of cost and of yield is greatest, and is always reached before the maximum sustainable yield. The tendency would be in an individually administered resource to cease hiring additional producers at the point E. Theoretically, it is in the interest of a country—and indeed of international society—not only to conserve the resource, but also to conserve the labour and capital employed in its exploitation. In practice, however, it is impossible to tell with precision when the point of maximum sustainable yield has been reached; the tendency is for the build-up of a chronic excess in catching capacity, and for equilibrium to be reached in a situation characterized by a large amount of effort, a depleted fish stock and a low sustainable yield. The fishery then tends to drag on precariously, with no margin to buffer the inevitable fluctuations. It has been pointed out, by contrast, that the only really satisfactory economic situation is found in the early phases of a new fishery, when there is expansion against a virgin stock.[3]

The above analysis traces the development of a fishery in total revenue and cost and in net profitability. It can apply fairly directly to particular fisheries in the modern period from the later nineteenth century, such as the North Sea plaice fishery and the distant-water trawl fishery for cod, and more recently to the Peruvian anchovetta, as well as to European exploitation of species like scallops and crab. Among the producers themselves, however, there will be differing degrees of efficiency, and it is also necessary to examine the situation of the marginal producer. He will be the one the value of whose catch equals his costs, which may (for simplicity) be assumed to be equal for all producers. The situation in Figure 1, in fact, indicates the progressive reduction of the least efficient producers as the numbers of fish available to each producer declines. Figure 2(a) shows the situation as the fishery develops towards equilibrium,[4] which will here be defined as the point where average revenue and average costs are equal. If costs are equal for all fishermen, the marginal and average costs will be equal, and the maximum net revenue per fisherman is given by the vertical distance of the average revenue curve above the point of intersection of the marginal revenue curve with the line of marginal cost (i.e. by EF). In this case the dimensions of the maximum net yield of the whole fishery will be given by the rectangle EFCD.

While it is obvious that such idealised models as these make a number of simplifying assumptions, they can be employed to show the results of major changes in circumstances within the framework of a common property resource to which entry is free. Thus if fish prices were to rise (Figure 1(b)) the revenue curve would rise more steeply, and the equilibrium point would be reached at a higher total revenue (OG) and with more producers (OB) than before. The introduction of an improvement in fishing technique (Figure 1(c)) would also result in a more rapid rise in the curve, but in this case there would be a subsequent decline to an equilibrium with a lower revenue (OC) and a lower number of producers[5] (OB). It is also possible to apply the same methods to investigate situations in which several grounds (or stocks) are fished; while the equilibria are more complex, it is possible to compare (for example) the exploitation of grounds at different distances from a base, and of different levels of productivity. With greater distances from base, costs will rise more steeply, and hence the equilibria points for more distant grounds will be reached at lower levels of fishing intensity. On the more productive grounds, on the other hand, net costs will rise less steeply because of the greater revenues obtained, and the equilibrium point for such a ground will be reached with a relatively large number of producers. This latter has been in effect the modern situation of many of the Arctic grounds despite their distance from fisheries bases.

More complex changes could also be built into the same model: thus natural fluctuations in the fish stock would lead to an irregular curve of yield, while increases in the costs of boats or gear would mean that the cost graph would be non-linear. The equilibrium to which the fishery would tend, however, would again be that of a zero net yield.

The theoretical considerations described above have a direct applicability to several sectors of the European fisheries. While stocks very generally were until modern times exploited at levels below those of maximum sustainable yields, by 1900 some of the stocks of the southern North Sea—particularly the plaice—were being exploited beyond it. Since then the area in which exploitation of stocks beyond this point has been observed has extended outwards as increased pressure has been brought to bear on other grounds and stocks. It is now clear that the stock of Arctic cod in the north-east Atlantic is being exploited beyond its maximum sustainable yield, with adverse results for the profitability of the distant-water trawling industry of several nations which depend heavily on it. This has also had the result of bringing increased pressure on the cod stocks of Iceland, Greenland, Newfoundland and Labrador, and also on the redfish, haddock and plaice stocks of the Arctic, so that other fisheries also are in danger of being depleted to unprofitable levels. The haddock and plaice stocks of the North Sea are also exploited well beyond their maximum sustainable yield, and this has brought pressure on other stocks like whiting and sole.

The application of effective remedies to situations of over-fishing is very difficult. To date the main measures include (at the national level) the expropriation of part of the resources by extending fishing limits outward from the coast, and (at the international level) agreement on minimum mesh sizes to be used in different fisheries to allow bigger proportions of the stocks to grow to maturity before capture. While the extension of a nation's limits will help improve the profitability of the home fleet, it will act adversely on the efficiency of the inter-national fleets by restricting the area in which there is free competition; extension of fishing limits out to twelve miles from headland-to-headland base-lines, however, is now usual around the North Atlantic. Even within national limits, however, efficiency is frequently lowered by the reservation of the stocks for smaller inshore craft at the expense of vessels like bigger trawlers: this occurs, for example, in Norway, Iceland and Britain. The regulation of mesh sizes in nets has been shown to be an effective conservation measure, and has allowed an increase in yield of, for example, the haddock fishery around the Faroe Islands, at least in the short term. By agreement on the mesh sizes of nets it is possible to apply the principle of eumetric yield,[6] as formulated by Beverton and Holt: this holds the yield level (by weight) constant and also preserves the right of free entry; but the size of mesh is progressively increased so that an increasing proportion of the stock is allowed to grow to greater ages. While such a principle does allow flexibility, it also means a progressive reduction in the catch per unit of effort. Another conservation measure which has been advocated is the closure of nursery areas to fishing to allow stocks to mature; within the complex ecological systems of the sea, however, these may not be areally very distinct, and the nearest approach in this field has been the imposition of national limits within which foreign and bigger vessels are prohibited from operating: this does help conserve the young plaice stocks on the Dutch and Danish coasts, for example.

A measure used only to a limited extent in European fisheries is the imposition of seasonal closures or restrictions. Seasonal limitation of salmon and trout fisheries—especially in inland waters—is common; there are also some examples in open-sea fishing, such as the winter prohibition of prawn-trawling in the Scottish Moray Firth, and the limitation on the use of the purse-seine in the Lofoten cod fishery to months outside the main season. Short-term restrictions on herring landings have also operated in several countries in times of glut, and in Scotland the Firth of Clyde is closed to herring trawling for part of the winter season. Most of these seasonal limitations on effort in Europe, however, have until recently represented political measures to protect the interests of small producers rather than to conserve the resources. An agreement of the North-East Atlantic Fisheries Commission, however, reached in 1970 as a result of serious disquiet about herring stocks, will enforce closed periods in the North Sea for this important species in May and from 20th August to 30th September. This will inhibit the catching of poorer quality fish at the beginning and end of the main herring season. There are, in fact, better examples of seasonal limits on effort in North America, with the North Pacific Convention on halibut, and seasonal closures on sea fisheries for lobster and salmon.

Seasonal closures are not, however, economically desirable, since they render both labour and capital partially inactive.

While all the conservation measures mentioned above help to perpetuate the resource, they have all a major economic weakness in that they do nothing to restrict free entry into the industry. To regulate entry in a competitive international situation would certainly be very difficult, although in the modern situation entry is not entirely uncontrolled. While no European country has gone as far as Canada in licensing all entry into some fisheries, the fact that building of fishing vessels is now subject to government aid in most countries does mean that capital, competence and experience of companies or boat-owners are demanded; and in eastern Europe the size of the fleets is planned on a national scale. In Norway, trawlers are limited by license, largely to protect the interests of smaller fishermen (this compares with the distant-water fisheries of Japan, which are the world's biggest and entry into which is completely controlled by licensing). With the move towards a greater measure of uniformity in modern vessels and techniques, it would theoretically be possible to compute an optimum economic level for the fleets exploiting different stocks and control the entry into each accordingly.

Fisheries very generally have special systems of their own for paying the labour employed, and these have generally involved the allocation of a particular proportion of the proceeds to crew members, irrespective of time worked or other conditions. It has been observed that extreme profit-sharing distinguishes fishing from all other industries.[7] (This has generally been the case in Europe, although it is less pronounced in some countries—notably West Germany, the Netherlands and Ireland). While it is theoretically possible to compute the optimum crew size in these circumstances as the number which gives the maximum average production per share, the arrangements made by small operators are often still recorded with little precision, and it is often difficult to estimate the real cost of labour, and to compare it with other employment. Another complication which is common even in the most advanced countries is that much of the labour for inshore fisheries is provided on a seasonal, part-time or casual basis, which necessarily limits any statistical analysis.

A characteristic of much of the labour in the more traditional fisheries is that it is widely dispersed along great lengths of coasts—shown above all in Europe by Norway, but known in Italy, around the Baltic, Brittany, and elsewhere. Traditionally, such labour has been organized within small units in community or even family groups, has shown restricted mobility both areally and occupationally, and has been slow to react to changing conditions. Because of this there has often been an excess of labour applied in the conditions of free entry to fisheries. This has often led to demands for subsidies from government funds, and state aid of some kind is now usual in nearly all European countries. The dispersed and often isolated location of much fishing manpower has also meant that it has proved difficult to find alternative employment without migration; and other industries have shown little tendency to move to fishing settlements, even when national policies have been adopted of giving incentives for them to move to peripheral locations.

In the countries of western Europe labour in fishing communities is now becoming more mobile, and in the bigger scale fisheries like distant-water trawling and purse-seining it can generally be more accurately costed. Further, the cost of labour in proportion to that of capital is decreasing, often at a rapid rate. The tendency now is for the payment of a basic wage (on a daily, weekly or monthly basis) together with a share of the proceeds. For such fisheries, profitability can be more readily assessed, and points of marginal productivity defined. The tendency is for the proceeds per share (or the total proceeds per man) to be lifted to a point determined by the opportunity cost of labour in other industries; indeed, with the uncongenial nature of much fishing employment it may have to go considerably above this level if labour is to be retained. This is the experience of distant-water trawling in western Europe, and the situation appears to be paralleled in eastern Europe. While income levels are generally good, at least in the capital-intensive fisheries such as British and German distant-water trawling, they are actually low in relation to labour inputs, as long hours are frequent.[8]

This greater mobility of labour is fostered by the increased proportion of operation from bigger harbours and ports, rather than from isolated settlements, and the central location of most of the more sophisticated fish processing plants enhances this concentration. In major urban centres there is more alternative employment and a more competitive labour market.

In the developing world, the considerations of labour can be very different, as there can be a great premium in providing work for as many as possible while capital may be in very short supply. In these conditions, the prime consideration may be the increase of food supplies in which flesh or protein foods are of great importance, while labour here is generally of low cost. Such considerations do not apply in Europe on anything like the scale of south-east Asia, Africa or Latin America, but do

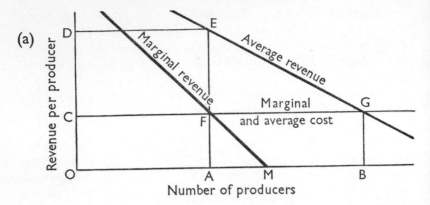

(a)

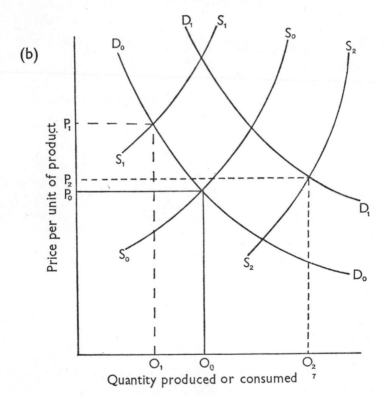

(b)

Figure 2.

(a) Curves of average and marginal cost.
(b) Supply and demand curves.

have some relevance in the poorer countries of the Mediterranean and eastern Europe.

The economic relationships in the marketing and distribution of fish are less unusual than on the catching side, although this sector has its own special characteristics. Since it has been less subject to economic distress in recent times, it has been much less the subject of government support, and as a result comprehensive statistics are rare. Data on consumption have often been derived indirectly from national differences between landings, exports and imports, and any comparative analysis is limited by the different systems of classifying and recording data. Some surveys do exist of distribution, but these have been spasmodic.

Obviously, demand for fish is a major factor influencing fishing intensity, and the relation between the two can be shown by simple supply and demand curves.[9] In Figure 2b, the catch or output level is reached where the two curves intersect (O_0) and at the equilibrium price P_0. Increased exploitation would reduce the stock and hence raise both costs of catching and price; the supply curve would shift to S_1S_1, and in the new equilibrium situation the catch would be O_1 and the price P_1; this would be reached when the yield dropped to a level at which it could be maintained with a less dense stock. While this sort of situation is known in modern Europe, especially in shell fisheries, more frequent is the situation by which advances in technology increase the supply (e.g. to S_2S_2). This decreases price levels and induces increased consumption; the demand curve will shift to D_1D_1, and O_2 and P_2 will represent new equilibrium levels of output and price. While the latter would appear broadly to represent the situation in eastern Europe, and in the Mediterranean, in the advanced countries of western Europe, the situation is more complex, as the tendency is for technology to advance but for supply to remain fairly constant in the long run, due both to decreases in fleet size and to stocks being exploited beyond their maximum sustainable yield.

The marketing and distribution sector has two main branches; on the one hand that in which fish bought at a coastal market are sent direct for consumption; on the other, the branch in which processing—and hence an extra cost increment—occurs before the dispatch to the consumer. The latter branch is tending to become dominant in Europe, and the tendency is for the share of the more sophisticated processors to increase. Direct delivery for consumption, however, still plays a big part, not only in the less developed countries of the Mediterranean, but also in countries like Britain, Denmark and Belgium, where big populations are within a short distance of the coast. This sector is naturally less important in the big fish exporters like Norway and Iceland, and among the eastern European countries the fleets of which operate at long range and do much processing at sea.

The specialized and perishable nature of fish means that distribution costs are both high and variable; and a wide margin between the price on the quay and the cost of the final produce to the consumer is characteristic. While distance between port and consumer is a factor influencing the size of this margin, processing costs, the number of links in the distribution chain, and the measures necessary to minimize spoilage are normally greater cost factors. Another disadvantage to the fisherman is that small changes in demand tend to have magnified effects at the producer's end of the system, as most of the intervening costs are fixed and unaffected by fluctuations.[10] Investigations in the U.S.A. have shown that the effect of market conditions on producers' income is more marked than in almost any other sector of the national economy.[11]

In the marketing of fish, the usual situation is that a distinct market exists for each species, even although a range of species is handled by port merchants and by successive links in the distribution chains. A limited degree of substitution does occur, however—such as whiting for haddock in times of scarcity, and various other flat fish for plaice; but the market allows very little substitution for such important species as herring, cod, sardine and tuna, and this applies even more to high value products like salmon, lobster and scallops.

Demand for fish shows a notable distinction between developed and developing countries. In the highly developed countries of western Europe, the general trend is for per capita consumption to remain steady or go down, and for demand elasticities to be low, so that rises in per capita income lead to little increase in fish demand. In these circumstances, total demand tends to increase only as a result of increase in population. The tendency in eastern and southern Europe, however, appears to be for an increase in per capita consumption, and for higher elasticities of demand. Total demand is also now strongly influenced by the extensive use of the products of reduction, especially in animal feeding stuffs. There has been a very rapid recent increase in this demand from the later 'fifties; it is associated especially with the intensification of stock farming, and the main location in Europe is the more highly developed countries of the west. This sector is so recent that it is as yet difficult to discern an equilibrium, so that detailed economic analysis of the situation is not possible. The main species fished for reduction in Europe are herring and sprat, although surplus and inferior stocks of other species may be used. The marketing of the products of reduction—and hence the fisheries themselves—is now greatly influenced by the Peruvian anchovetta fisheries for reduction, which now contribute the bulk of the world supply. There may be a series of levels in the market for each species, especially for those used for reduction, and this is best shown by herring. Here the biggest prices are realized for filleting, canning and for whole gutted fish for the fresh market; lower steps are represented by simpler processes like kippering, marinating and curing; and use for reduction realizes the lowest prices. Demand is thus effectively 'stepped' and the lower price users tend to have more erratic supplies.

The usual situation in fisheries is that there are numerous independent catchers who are the suppliers, and this generally means that they are unable to influence the marketing process, as concerted action is unusual, although in Norway, fishermen's unions do have programmes of action with regard to supply. In the effort to give more stable economic conditions, fixed or minimum prices may be fixed either by agreement of the interests at particular ports, or by government action at a national level; these can, however, detract from overall economic efficiency, and may even operate to the detriment of the fishermen themselves.

Fish consumption must, of course, be seen within the framework of the general food market, in which its most direct competitors are other flesh and protein foods. In the advanced countries, total per capita food production is fairly static, and as income increases demand elasticities for foodstuffs tend towards zero. There are changes, however, in the proportions of different items in the diet, with the substitution of more preferred items for less. Fish is one of the less preferred foods, and never accounts for more than 3 per cent. of the total calorie intake in a country; available figures in Europe suggest it commonly accounts for around 10 per cent. of the consumption of animal protein, although in some countries with a long fisheries tradition—such as Iceland, Norway and Portugal—this proportion may be multiplied several times.[12] While data are generally insufficient for the assessment of the cross elasticities of demand between fish and other sources of animal protein, the trend in advanced countries is for increasing per capita consumption of meat, poultry and eggs, and decreasing per capita consumption of fish. Fish have a high elasticity of demand, and this has inhibited the passing on of increased costs to the consumer. Nonetheless, it appears to have an assured place in the diet and on the market in advanced countries, because of the element of variety it provides. It is noteworthy, too, that the per capita consumption of the higher value fishery products like shellfish, salmon, frozen fillets and pre-cooked fish has been increasing. In lower income countries, there is a marked increase in demand for fish with raising of income; the trend is away from a diet of excessive carbo-hydrates, in the form of cereals and other starchy foods, and an increase in flesh food and proteins. This is the general situation in eastern and southern Europe, and this is enhanced by the lesser suitability of these parts of the continent for stock farming. Something of a similar increase in demand appears to occur in low income groups in advanced countries.

A factor which increases costs throughout the industry is short-term fluctuations in supply: capacity in both catching and processing is generally installed to cope with a level of supply considerably above average, and this increases the proportion of overhead costs. For that part of the fluctuations which is seasonal it is possible to take counter-measures, although costs must always be incurred above the levels that would obtain with a regular level of activity through the year.

Seasonal fluctuations are generally most marked in traditional inshore fisheries, but are found to some extent in all. Within the older, simpler economic structures, seasonality in the fisheries was seldom seen as an essential handicap, as time was always required for other tasks, especially in societies with a large element of subsistence activity, where much of what exchange occurred was at a local level. There was relatively little loss in having the simple capital equipment employed—small open boats, simple lines and nets—lying idle for part or most of the year: this has been the case with Norwegian cod fishermen, tuna fishermen in the Mediterranean, and salmon fishermen on the British coasts, for example. For the remainder of the year, men might have small-holdings to cultivate, the family's winter fuel supply to secure, and various domestic handicrafts to engage in; and they might get supplementary seasonal employment in such occupations as harvesting, logging or trapping. Even where bigger craft and more capital were employed, as was the case often with the herring fisheries, at least some of the crew might have employment outside the fishing season in making or maintaining their gear, and the boats might also be used for trading voyages.

With the increasingly capital-intensive character of the fisheries, there is a strong tendency to eliminate the seasonal laying up of vessels. Basically, this can be achieved either by greater mobility, or by seasonal conversion of vessels to another fishery which is locally available. Added mobility is probably best instanced with the modern fleets of purse-net herring boats, most of which aim to operate the year round. The big Norwegian fleet, for example, built in the first instance to exploit the home winter herring fishery, range in summer into the North Sea, to Iceland and the Arctic to find herring shoals. In Britain the herring fleets developed a considerable range in the latter nineteenth century, when drift-net vessels moved all around the British coasts, and this is now continued with modern fleets employing the pair-trawl and purse-net. While such year-round fisheries improve the economic position of the catching fleets, they tend to put a strain on shore processing and transport, which need to be geared to seasonal fluctuations on different coasts. Distant-water trawler fleets are essentially mobile, and may range over different Arctic grounds with the season. The difficulty here tends to be with an excess of supply in summer, when fishing is less interrupted and when white fish demand falls off seasonally; and this can lead to laying up sections of the fleet.

Seasonal conversion to other fisheries is fairly common, but it tends to become increasingly difficult the larger the vessel employed; the time and expense taken to adapt large vessels in the short term is proportionately greater. On many coasts, small inshore fishermen have a range of gear which can be used for different species, and the time taken to switch from lobster to mackerel fishing, for example, may be insignificant. On most of the Norwegian coast, a pattern which includes one major fishery together with supplementary ones at the other seasons has been characteristic. Thus crews who engaged in the winter herring fishery might fish mackerel and sprat at other seasons, the northern cod fishermen could catch saithe and herring in summer, while vessels which spent most of the year in long-lining might fish small whales in summer. In Scotland, too, the traditional drift-net herring fishermen went long-lining for white fish during the lean spring season, and their modern counterparts may engage at the same season in seining for white fish. Most of the vessels now converted seasonally to different fisheries are less than 80 feet in length, but there are examples of conversion among bigger craft. Thus in the Faroe Islands, a number of modern vessels of around 120 feet have been built for long-lining for white fish and purse-seining for herring; in Iceland vessels of a comparable size are used for the same purposes, and also for ground-seining and gill-netting in the cod season; and even bigger vessels in the eastern European fleets may alternate between white and herring fishing: they may, for example, exploit the cod stocks at Newfoundland in summer, and the winter herring off the Norwegian coast. Even with such adjustments, however, some seasonal fluctuations in the effectiveness of the deployment of capital and labour are inescapable.

The operation of processing plants is generally more forcibly affected by short-run and seasonal fluctuations in supply than are the fishing fleets. The result tends to be increased overhead costs in excess processing plant and in storage space, together with increasing difficulty in recruiting a seasonal labour force.

This is partly because shore-located processing plants are intrinsically immobile, while the fleets supplying them range over varying distances. While factory ships do exist, the very much greater cost of installation and operation of plant aboard ship in most situations more than cancels the advantage of mobility. For the U.S.S.R., Poland and East Germany, however, which have big fleets operating at ranges of several thousand miles in both the North and South Atlantic, the relative cost of processing plant aboard ship is lower, and floating factories are an essential part of their organisation. Some smoothing of the flow of supply to individual plants may now be achieved by the use of radio to direct vessels away from points of glut to those of deficit. In traditional fisheries, seasonal closures of plant were very general, but it was not unknown for labour to move to similar plants in different locations with the seasons: this was very characteristic with herring curing in Britain until the Second World War.

The adverse effects of short-run variations in supply are greatest in pelagic fisheries, as both day-to-day and seasonal fluctuations tend to be of higher amplitude; in addition, storage is often more difficult as the bulk of landings is often greater, while the more oily pelagic fish deteriorate more rapidly. The effect of the 'stepped' demand curve especially characteristic of pelagic fisheries has been indicated (p. 23), with the resulting excessive variations in supply for low-value users. In white fish processing, whole-year operation is usual, although there may be seasonal variations in the tempo of activity. Generally processors supplied by distant-water trawling fleets have the most regular supply, although even here winter landings are lower. Where the supply comes wholly or in part from inshore vessels, a greater amplitude of fluctuation is characteristic. In both Iceland and Norway, the main cod-fishing season is in the early months of the year, up to April or May. In Norway, especially, a considerable part of the catch comes from small inshore vessels, and here the month-to-month landings over the year can vary by a factor of ten.[13] While extra storage space and plant is required for the peak season, a considerable levelling out in employment can be achieved by minimizing processing in February and March by concentrating on block fillets.[14] In the remainder of the year more refined processing, including pre-cooking can be practised, although this does involve further capital expenditure on plant which is seasonally idle. Seasonal variations are also characteristic of salmon and shell fisheries, and their processing and marketing. In Scotland, for example, seasonal closures are characteristic in crab processing plants, while merchants dealing in salmon and trout must turn to white fish.

The possibilities and costs of storage may vary seasonally; in Europe generally, rates of spoilage are higher in summer and greater icing or larger cold stores may then be necessary. In the Norwegian winter fisheries low temperatures considerably inhibit spoilage and, while it is possible to store herring for reduction up to three weeks at this season, one week is the maximum in summer.

For the processors, the effects of seasonal fluctuations may be minimized by transporting supplies from distant points in off-seasons for local landings, although this necessarily adds an increment to costs. Thus herring processors on the east coast of Britain, whose main local supplies are in summer, may operate in winter with supplies brought by road from the west coast, or even by ship from Norway, and German processors depend at this season on the same sources. Norwegian reduction plants may occasionally use sprat catches from Scotland, while plants on the east coast of Scotland cooking and freezing crabs and nephrops may bring supplies from the west coast in winter to balance their home summer landings.

While it is possible to make some large-scale comparisons between fisheries enterprises—especially between countries, measurement of comparative advantage can seldom be approached. The size of the fishing industry of each country is related to the cost of labour and of capital in relation to the value of the catch. It is also related to the opportunities for the deployment of labour and capital in other fields, and the availability of alternative sources of flesh food and protein.

The comparison in cost of boats and gear which are internationally traded is possible, and this tends to apply increasingly to the bigger vessels and more complex gear, although there are sectors of both (smaller craft and simpler gear) in which markets are still substantially contained within national frontiers. Labour, however, is generally difficult to cost; and markets are complex in that the fresh fish sector is substantially supplied by national fleets only: this occurs behind the partial protection of tariff walls, and, although there is increasing international trade in processed varieties, the volume and direction of flow may also be modified by tariffs. Trade within the Common Market countries is now moving in less obstructed channels, and any enlargement of its membership would extend the area within which this occurs.

In the open sea fisheries outside national limits a very complex situation of competition often occurs. Even when prosecuting the same fishery, fleets of different countries have different cost structures in their operations, and different price structures in their markets. Often the proportions and costs of capital and labour will be different, and their distances from base and time at sea will vary. It is also possible for distinct fleets to be exploiting different species or sizes of fish in the same area. In many such situations of international competition, meaningful comparisons are almost impossible. The tendency now is for an increasing development of capital-intensive fisheries and for a greater international flow of trade in vessels and gear: this could make such comparison less difficult in the future.

REFERENCES

1 H. Scott Gordon, 'The Economic Theory of a Common Property Resource: The Fishery', *Journ. of Pol. Econ.*, 62, 1954, p. 136.

2 F. T. Christy (Jr.) and A. Scott, *The Common Wealth in Ocean Fisheries*, 1965, pp. 6–16.

3 J. A. Crutchfield, 'Economic Objectives', in *The Fisheries, Problems in Resource Management*, ed. J. A. Crutchfield, 1965, pp. 43–51.

4 H. Scott Gordon, *Ibid.*, p. 130.

5 F. T. Christy (Jr.) and A. Scott, *Ibid.*, pp. 12, 13.

6 R. J. H. Beverton and S. J. Holt, 'The Theory of Fishing', in *Sea Fisheries. Their Investigation in the United Kingdom*, ed. M. Graham, 1956, pp. 413–16.

7 H. Zoeteweij, 'Fishermen's Remuneration', in *The Economics of Fisheries*, eds. R. Turvey and J. Wiseman, 1957, pp. 18–19.

8 C. Campleman, 'Manpower Problems in Fisheries', in *Some Aspects of Fisheries Economics: Report of a Research Seminar held in Bergen, Norway, 28th Jan.–22nd Feb 1963*, ed. G. M. Gerhardsen, 1964, II, p. 41.

9 F. T. Christy (Jr.) and A. Scott, *Ibid.*, pp. 82–84.

10 J. A. Crutchfield, *Ibid.*, p. 44.

11 F. W. Bell and J. E. Hazleton, *Recent Developments and Research in Fisheries Economics*, 1967, p. 7.

12 F. T. Christy (Jr.) and A. Scott, *Ibid.*, pp. 19–20.

13 P. A. Flaate, 'The Policy of the Firm from the Fleet Manager's Point of View', in *Some Aspects of Fisheries Economics*, ed. G. M. Gerhardsen, 1964, III, p. 10.

14 H. Henriksen, 'Economic Aspects of Fish Processing, in *Some Aspects of Fisheries Economics*, ed. G. M. Gerhardsen, 1964, III, p. 58.

CHAPTER III

The Resource Base of European Fisheries

The volume—and in general terms the value—of fisheries resources is determined in the first place by factors of marine ecology; these determine the location of different species, the size of the stocks, and to a large extent the variations in them in both place and time. The size of the resources, however, is now increasingly influenced by man himself, as modern catching rates have altered the natural balance of many of the ecosystems of the sea in considerable degree.

The North Atlantic may be divided into two zones on the basis both of fish species and of productivity: the more productive zone is in the Cool Temperate and Arctic zones and is approximately bounded on the south by the annual marine isotherm of 13°C.[1] (Figure 3). This line runs approximately from New England to Brittany. There is thus a major contrast between sea and land in the abundance of life in different latitudes, the maxima of the latter being in the Equatorial Zone. This more productive zone is considerably wider in the north-east than the north-west of the Atlantic, and this is related to the basic anti-clockwise circulation pattern in this part of the ocean. The flow of the warm Atlantic Drift into the north-east part of the basin is felt as far north as lat. 78°N. off Spitzbergen and conditions suitable for fish together with open water can be found here all year. In the western part, however, the southward flow of the East Greenland and Labrador currents brings pack-ice southward; in this part of the ocean fisheries do not extend into the Arctic circle, and in winter and spring the pack ice can extend south beyond Newfoundland into lat. 43°N., which is near the southern limit of the zone. North of the 13°C annual isotherm there is a wide range of commercial species, and they include the supremely important cod and herring. To the south of the

13°C. annual marine isotherm, the waters of the North Atlantic are considerably poorer, but still provide important fishing grounds. The main commercial grounds of both these

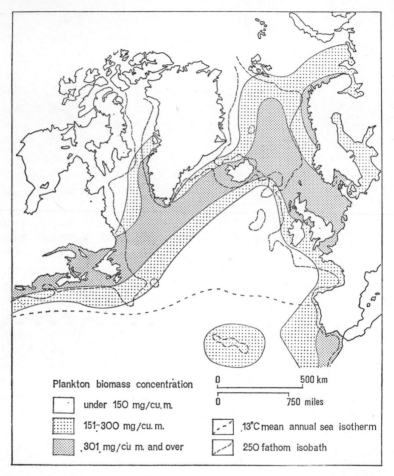

Figure 3. Main physical factors influencing fisheries productivity in the North Atlantic.

sectors are on the marginal continental shelves, although the area of these is much greater in the north. The almost enclosed waters of the Baltic, Mediterranean and Black Seas have each distinct complexes of fishery resources, but they are relatively poor; the Sea of Azov and the Caspian Sea, however, both show very high levels of productivity. On the continental scale, other inland water bodies are of limited significance.

PLANKTON

The variations in distribution and abundance of fisheries are determined by the dimensions of the plankton 'crop' (the basic foodstuff) in different waters, and on the structures of the food chains and food pyramids that ultimately determine what fraction of the 'crop' may be harvested as fish. Life in the sea, as on the land, is basically dependent on vegetable matter, and the vegetation of the sea are the minute phyto-plankton. As with other vegetation, variations in temperature and insolation are important in determining regional variations in the phyto-plankton; but while availability of moisture tends to dominate among other factors in land vegetation, the crucial factor in the sea is most frequently the availability of certain key nutrients, particularly phosphates and nitrates. Other salts which occur in greater amounts in living matter, such as sodium chloride and calcium carbonate, are virtually always present in adequate quantity; but the key nutrients may be present in such dilution that they set low ceilings on the development of phyto-plankton. The abundance of key nutrients is largely explained by the three-dimensional circulation of the ocean waters: vertical movements within or at the junctions of water masses are involved as well as horizontal currents. In the richer waters of the North Atlantic, there is little or no permanent temperature gradient between the surface and deeper waters, and phosphates and nitrates are periodically renewed from below by the stirring action of winds and currents. Greater solar heating of the waters further south, however, results in the surface waters being at a temperature of at least 10°C. above the deeper layers, and are separated from them by a permanent thermocline. In these conditions there is little renewal from below, the nutrients being cycled within the surface layers. There is an exception to this off the Atlantic coast of Africa, where the predominant off-shore movement of air masses results also in an off-shore water movement which is balanced by upwelling from the deep ocean basin. Similar thermoclines in the Mediterranean and Black Seas are largely responsible for their poorer resources.

The main period for the renewal of key nutrients is the winter, when wind strengths are greater, but this is not reflected in plankton numbers until the rising temperatures of spring lead to a burst of activity. In Arctic waters there is a marked single summer peak in phyto-plankton but in the temperate zone the growing season is longer and the peak less marked. In parts of the temperate zone, including the northern North Sea and the Celtic Sea, double peaks are observed: the spring flowering of the phyto-plankton is followed by a phase in which available nutrients are used up while a seasonal thermocline develops in the water, and the phyto-plankton are grazed down by the zoo-plankton; subsequently, autumn gales cause vertical mixing with renewal of nutrients, and a lower second peak of plankton numbers is observed before they are reduced by the lower temperatures of winter. The zoo-plankton forms, which are generally larger than phyto-plankton and live longer, show considerably less annual fluctuation, and some species are a store of winter food for fish.

The relative organic wealth of different sea areas can be shown by the quantitative estimates that have been made of biomass (i.e. mass of living matter). Planktonic organisms may average over 400 million cells/cu.m.,[2] and they can never all be included in a sample obtained at sea as some will pass through the finest mesh tow net. However, comparative estimates are available for different areas, and in the most fertile areas of the Atlantic, the mean plankton biomass over the year exceeds 300 mg./cu.m., although seasonal values of 2,000 mg./cu.m. may be obtained. The highest values are found within the boreal water (Figure 3): a zone in which mean annual values exceed 300 mg./cu.m. forms a narrow coastal strip along the eastern seaboard of the U.S.A., and broadens out north-eastwards to include Newfoundland and Iceland; it also includes the North Sea and the great part of the Norwegian Sea. On either side of this zone are narrower parallel zones in which the concentration lies between 150 and 300 mg./cu.m., and the values in a coastal zone from western Ireland to Portugal also have this range, as does the Azores Plateau against which there is upwelling from the deep ocean basin.[3] Deep upwelling is also responsible for the high values in excess of 300 mg./cu.m. that extend in a coastal zone from south Portugal to Sierra Leone: this is flanked to the west by a continuation of the zone of 150 to 300 mg./cu.m. that extends south in a tongue that crosses the equator. While the Mediterranean, Black and Caspian Seas all have lower values of standing plankton crop than the most fertile areas of the open ocean, in them the rate of grazing down of plankton is higher in warmer waters; the result is that the fertile Sea of Azov and the Caspian Sea, in which only seasonally do plankton biomass values reach 150 mg./cu.m., support greater densities of fish stocks than any part of the ocean.

A major factor governing the availability and composition of the fish stocks is water depth. It has been estimated that the relative productivity of different sea areas may vary by a factor of 50 or more;[4] and, on the world scale, the total plankton biomass on the continental shelves is twice that of the deep oceans, despite their much smaller total area. This is due both to greater availability of nutrients and to their more rapid re-cycling. When the sea floor is within the euphotic zone (i.e. within the range of light penetration) it supports a much richer ecological community, which includes a variety of benthos (organisms of the sea floor). Demersal and shell fish are in food chains and food pyramids which include organisms of the sea floor, and the fisheries for them are located on the continental shelves, off-shore banks and (more rarely) on the topmost reaches of the continental shelf at the edge of the ocean deeps. In the North Atlantic, the continental shelf is best developed in the north-east, especially around the British Isles, and is also extensive around Iceland and Spitzbergen. In the north-west Atlantic, some of the richest fisheries of the world are located on the Grand Banks of Newfoundland. Most of the main pelagic fish stocks spend at least part of their life-cycles on the continental shelves, where the plankton crop is richer. Overall, the shelf areas are by far the most important areas fished by the European nations.

Among the near-enclosed and enclosed seas, the Baltic is shallow but productivity is somewhat limited by restricted water circulation and lower nutrient levels. The fresher waters, however, here support a distinctive assemblage of fish stocks.

Overall productivity of the Mediterranean and Black Seas is restricted by the fact that deep basins account for the great part of their area, but shallower sections like the Adriatic and the north-west Black Sea have much richer fisheries. The Sea of Azov and (to a large extent) the Caspian Sea are shallow, have high temperatures for much of the year and have nutrients recharged by big inflowing rivers: the productivity of these seas is the highest of the whole continent.

A good general index of areal variation in potential of the shelves is given by the estimation of benthos biomass.[5] In the ocean deeps, it is thought that the benthos biomass scarcely rises above 1 or 2 gm./sq.m., but in the deep Tyrrhenian Basin of the Western Mediterranean it rises to 6 gm./sq.m. Shallower waters have much higher figures: the Adriatic is around 183 gm./sq.m., the (shallow) northern Caspian 124 gm./sq.m., while the most fertile Sea of Azov has figures of over 700 gm./sq.m. Figures in the open sea for benthos biomass are on the whole lower, the general level for the continental shelves being from 50 to 100 gm./sq.m., although in the warmer conditions off southern Portugal, figures over 250 gm./sq.m. are usual.

FOOD CHAINS & FOOD PYRAMIDS

While the foundation of all marine life is the plankton, only a very small percentage of this is ever turned into commercial fish, even on the best fishing grounds. In this respect, fisheries resources compare very unfavourably with the feeding of crops of livestock on land, or even with open-range grazing. There is a heavy wastage in the food pyramids of the sea, especially at the lower levels where fish eggs and larvae, and plankton compete for survival and are subject to predation from higher forms of which fish are only a minority. The process is also rendered inefficient by the selective feeding of the commercial species, which are relatively limited in number, at all stages of their development; some of them actually prey on one another. At best the efficiency of conversion from any level in the population pyramid to the one above is only 10 per cent., and in some cases has been shown to be 5 per cent. or less.[6] The nutrition and growth cycles of the main commercial species of the North Atlantic are known, and it has been estimated that in the sea areas around the British Isles, for example, only about 0·05 per cent. of the yearly production of phyto-plankton is ever harvested as fish. While the fraction of benthos turned into fish food is higher, it is still small: observations in the Kattegat gave figures of 1 per cent. to 2 per cent., while in the southern North Sea 5 per cent. to 10 per cent. has been found. The numbers of links in the food chains have an important bearing on the fraction of the plankton crop that can be harvested. Only a very small fraction of the phyto-plankton crop is used directly by 'filter feeding' pelagic species such as the sardine. Pelagic fish feed mainly on the zoo-plankton, although their diet can include higher forms; but selective feeding limits the plankton fraction utilized, as is instanced in the strong preference for the copepod 'calanus' in the feeding of the important herring in the north-east Atlantic. The feeding of mature demersal fish is usually at least one stage removed from the plankton, and may be as many as three stages. Mature cod, for example, commonly have a diet two or three stages from the plankton, and this includes other commercial species like herring and capelin. As a general rule, fisheries operations are unlikely to be viable unless the standing crop of fish on the grounds is at least 8 to 50 gm./sq.m., depending on the species and on the economic efficiency of the fishery.[7] The overall average concentration for the North Sea—one of the best areas—is only 5 gm./sq.m.: this highlights the importance of more local factors in governing the distribution of the fisheries within the broad framework set by plankton and benthos biomass, and water depth.

FISH STOCKS & GROWTH RATES

The complex of factors which localize fisheries in particular areas include a number of varying scales of importance: temperature is a major factor, but salinity, weather, conditions of current and tide, on occasion drift ice and water turbidity, can all affect the location and intensity of operations. Plankton content and water depth, also, can operate at more local levels than indicated above, and bottom conditions may influence the available stocks. In a number of cases, the fish stocks have been shown to depend on particular water masses, or the inter-action of water-masses, with their own conditions of temperature, salinity and plankton content. The assemblage of factors defines minimum and optimum conditions for each species, and a map of species distributions would show a series of over-lapping zones in the sea; and fisheries generally yield a mixture of species, although one may be dominant. Saithe, haddock and other species, for example are commonly caught with cod, and mackerel with herring. Within the broad zones inhabited by individual species, it is usual to find a series of distinct stocks inhabiting particular areas. There are different plaice stocks, for example in the Baltic, the North Sea and Iceland; distinct cod stocks in the North and the Barents Seas, Iceland, West Greenland and off Newfoundland-Labrador; and within the main division of the herring into Atlanto-Scandinavian and North Sea stocks there are several sub-divisions. The import-ance of these distinct regional stocks is that depletion by inten-sive exploitation is not usually made good by immigration from other stocks but rather by the recovery of the regional one.

The rate of growth and the age of maturing in different stocks are basic in determining the time required for replace-ment of the numbers removed by fisheries operations. The sprat, for example, matures in two years, but has a maximum age of seven, while the halibut takes 15 years to mature although it can live for more than 50. The Arctic cod which matures in nine years can live to 25.[8] For the stock of the latter caught at Lofoten, it has been shown that it reaches a weight of 1 kg. in four years, 3 kg. in seven years and 7 kg. in ten years;[9] there-after, more of its diet goes into maintenance than into growth, and hence the fishery is more productive if it is caught by this time. In the north-west Atlantic, there is a difference in the growth rate and age of maturity between the stock off southern Newfoundland and on the southern Grand Banks, and that off Labrador, north-east Newfoundland and on the northern Grand Banks. The latter mature in five to six years, at lengths from 40–52 cm., while the former reach maturity at six to seven years at lengths from 60 to 80 cm., and hence are more productive.[10] There are also considerable differences among the herring stocks; those of the Atlanto-Scandinavian stock which is fished at the Norwegian coast mature at ages between five and nine years, and may live to over twenty; while the North Sea stocks mature at three or four years, and almost never reach an age over eleven.[11] Even where stocks are less distinct, areal variations in growth rates have been observed: in the south-eastern North Sea the growth of cod is more rapid than in the north-west, which is deeper and colder. The North Sea haddock also shows similar rapid and slow growth areas, together with a zone of an intermediate rate which stretches from the east coast of Scotland to the Shetland Isles.[12]

Virtually all catches are mixed, especially among the demersal species, which concentrate less in shoals. In the higher latitudes, cod very much predominate in the demersal catches, and the proportion averages over 90 per cent. on the Bear Island and Spitzbergen grounds, although it is only about 25 per cent. in the North Sea, about 10 per cent. on the west coast of Scotland and less than 1 per cent. in the English Channel. Hake has its greatest importance to the west of Ireland where it averages about 45 per cent. of the demersal catch, and skates and rays in the Irish Sea and English Channel where their proportion is around 30 per cent.[13]

THE INFLUENCE OF PARTICULAR FACTORS ON THE DISTRIBUTION AND DIMENSIONS OF PARTICULAR STOCKS

(a) Temperature

Temperature is by far the main factor determining the density differences between surface and underlying waters, and the relative abundance of plankton incident on this has already been indicated. Within the Arctic and Cool Temperate zones which have the greatest areas of high plankton concentrations occur the stocks of the outstandingly important gadoids (cod, haddock, whiting and saithe) and the herring, the redfish and the major part of the flatfish.

Temperature also in large measure defines the limits of tolerance of the physical conditions for each species; and its influence can also be discerned in different phases of life cycles.

Among the demersal fish, the most plentiful single species is the cod, and its absolute minimum level of tolerance is around $-2°C.$, although stocks of commercial size are generally limited

to waters above 2°C.;[14] this means in general it can be fished
almost as far north as ships can go without being hampered by
ice. Towards the equator, few are found south of waters at
10°C., although the absolute upper limit is about 20°C.[15] and
the main cod concentrations are in a narrower zone in which
the water temperature ranges from 2° to 5°C. These limits
locate the main cod fisheries north of 60°N. in the north-east
Atlantic, but between 40° and 50°N. on the north-west. While
it is unusual to find cod in numbers in temperatures below 2°C.
in the north-east Atlantic, they do extend into colder conditions
in the north-west Atlantic, especially in the feeding season, and
big stocks have been observed there in temperatures in the −2°
to 1°C. range in the zone of mixing of the Labrador Current
with the North Atlantic Drift.[16] Although food supplies draw
them into these sub-zero temperatures, there are considerable
physiological dangers as it becomes more difficult for them to
prevent the salt content in their blood getting to fatal levels.
The distribution of the related haddock largely overlaps that of
the cod, although its zone is somewhat narrower, and is found
in the warmer sector of the cod's range, the main concentrations
being observed in temperatures between 4° and 7°C.[17] The
range of sea temperatures associated with herring in the open
sea is closely comparable with that of cod; but there are major
concentrations in the southern part of the zone.[18]

In the warmer waters off southern Europe there is found a
different assemblage of stocks which only in limited degree
overlap those of the north. They include species like octopus,
squid and cuttlefish, but the most important stocks here are
pelagic, including especially the sardine, anchovy, and the tuna.
The sardine is found mainly in the eastern Atlantic in temperatures between 10° and 20°C.;[19] this defines its distribution as
the waters from south-west England to north-west Africa, and
the Mediterranean. The tuna is the most widespread commercial species in the world's oceans, although it is less important to the European nations than to countries like Japan and
the U.S.A. It occurs in several species, of which the bluefin,
albacore and bonito are fished from Europe. The bluefin has
the widest temperature tolerance, and is recorded in temperatures from 5°C. to 29°C., while the albacore ranges from 14° to
32°C.[20] The main location of exploitations are Atlantic waters

off Iberia, but the bonito and albacore are both widely fished in the Mediterranean.

There are several commercial species the temperature tolerance of which extends from the boreal into the warmer southern waters, and these include white, pelagic and shell fish. Most important is the hake, which is found on both sides of the Atlantic, and is also exploited to some extent by European fleets in the Southern Atlantic off South Africa. It is found in temperatures from 8°C. to over 20°C.,[21] and the main grounds worked are in a zone from off the west coast of Britain to northwest Africa, while it is also found in the Mediterranean. Mackerel and sprat in the pelagics, and lobster, shrimp and scallop are also found in a broader temperature zone.

As well as setting broad areal limits to the occurrence of species, temperature influences the location and abundance of fishery resources in more particular ways: tolerable and optimum temperatures for spawning, egg development and feeding are usually set within narrower limits.

The north Atlantic mackerel is a species which spawns in temperatures between 10°C. and 15°C.; spawning in the Mediterranean is between January and March, but further north it becomes later, being delayed until the warming of the sea in spring and summer. In the Sea of Marmora mackerel spawn in April and May, between March and July to the south of Ireland, and in June and July in the northern North Sea.[22] Observations on Arctic cod stocks show a narrower temperature (3°C. to 6°C.) range for spawning, which mainly occurs in spring and early summer.[23] Off North Norway, spawning takes place in the early months of the year at around 6°C., at a depth defined by a seasonal thermocline; and on the opposite side of the Atlantic, less complete records suggest that off Newfoundland spawning occurs at about the same period in temperatures of around 4°C.[24] Plaice is another winter-spawning species, and prefers temperatures in the range 4° to 7°C.; this gives a continuous spawning period in the open sea (e.g. December-February in the southern North Sea), but spawning is interrupted and prolonged for the Baltic plaice stock by cooling of the waters to around 0°C. in winter.[25] The herring is less circumscribed in its spawning temperatures: herring of the Atlanto-Scandinavian stock spawn off the Norwegian coast in the early months of the year in waters at around 7°C., while in the North Sea in summer spawning may occur in temperatures of 15°C. The concentration of stocks in shoals for spawning is one of the most important factors for fisheries. Temperature also has its effects on other phases of the life cycle, although these are less precisely known. It has been shown, for example, that the optimum temperature range for the development of the eggs of the plaice is 15°C. to 17°C.,[26] and cod prefer temperatures between 1°C. and 2°C. at feeding times.[27] The optimum growth conditions for capelin occur in temperatures just above 0°C., while those for mackerel are around 10°C.[28] Temperature also strongly conditions the main feeding seasons with the annual fluctuations of the plankton crop, and in all the boreal water zone the main feeding times are marked in the warmer half of the year, and this is generally the case in southern waters without being so prominent.

(b) *Salinity*

Salinity is a minor factor in determining the location of commercial fisheries. The amplitude of variation within the open sea is very generally restricted to the range from 32°/oo to 37°/oo, which is well within the tolerance of most species, although their optimum may be within a more restricted range. The cod, for example, prefers values between 32°/oo and 35°/oo,[29] although it is found in the Baltic associated with values of 10°/oo. In the enclosed and near-enclosed seas, salinity has a notable effect, but this is qualitative rather than quantitative: the composition of the plankton crop is altered, and with it the structure of the dependent food chains and food pyramids. This explains the presence in the low salinity waters of the Baltic of a species like the pike (more usually associated with fresh water), and also the absence from the same sea of the mackerel for which the minimum tolerable salinity is 20°/oo.[30] The Baltic, however, has its own stocks of species like the cod, herring and plaice, and the Baltic plaice can tolerate waters of salinity as low as 7°/oo[31] although it migrates into the most saline water available (in the south-west) for spawning. The successful transplantation by scientists of the U.S.S.R. of the Baltic herring to the more saline Caspian is an indication of the adaptability of many species to different conditions.[32] Anadromous species like the salmon, trout and sturgeon (i.e. those which spend different parts of their life cycles in fresh and salt water) give special examples of adaptability: fisheries for them are found both in the sea and in rivers.

(c) *Sea Depth & Bottom Conditions*

The location of the main fisheries on the continental shelves because of plankton concentrations and the ecological link between demersal species and bottom-living invertebrates has already been indicated. Although pelagic species may range into deeper waters, they also spend at least part of their life cycles on the shelves in feeding and spawning migrations. While some herring stocks spend part of the year in the deeper waters between Iceland and Spitzbergen, they migrate on to shallower bottom for spawning: the herring is unusual in laying its eggs on the sea floor, and it characteristically spawns on gravel bottom.[33] The tuna also, although found throughout the warmer ocean waters of the globe, also migrates into shallow water to spawn, and in the Mediterranean and on the Atlantic coast of Iberia the main fisheries for it are inshore. In the Mediterranean as a whole, the great part of which consists of a series of deep basins, pelagic species are the most important. As well as being related to the restricted area of shallow water, this is due to the temperature regime being beyond the tolerance of the supremely important gadoid family (cod, haddock, etc.). While pelagic species do spend part of the year on the sea bottom, and commonly go down to the bottom during the day at other seasons, they are very generally caught when they rise to the upper levels of the water at night, as this has proved easier and more productive with both traditional and modern fishing gears. In the Mediterranean especially, lights may be used to attract the fish into the gear set. In the open Atlantic, the redfish (or rosefish) is a species of deeper waters, and its main distribution is in the north-western Atlantic,[34] more remote from Europe.

On a more local level, depth and bottom texture restrict the range of particular species. Anchovy are a shallow-water species found around the coasts of Europe from the North Sea round to the Black Sea, and are especially common in estuarine situations like the Humber, Wash, Ijssel Meer, and most especially the Sea of Azov.[35] Flatfish are usually found on soft sandy or muddy bottom in shallower waters. They are a main stock in the shallow and prolific Sea of Azov,[36] and a well investigated species is the North Sea plaice which is rarely found in depths of over 40 fathoms, and its main location is on the soft bottoms of the south-east. (It may be noted, however, that the Arctic plaice species has been found down to depths of 100 fathoms in the Barents Sea.[37]) In the deeper northern North Sea, the main commercial demersal species becomes the haddock, which is found associated with a greater variety of bottom texture than the plaice, which is the main demersal species in the south. Also more common in the north is the whiting, which is particularly associated with soft bottom. Halibut and ling are mainly species of relatively deep water, and some of the main grounds for them are found near the continental edge. When halibut come closer inshore, they often occur in the vicinity of rocky parts of the sea-bed, especially where these confine current streams, and the saithe also is commonly found in these conditions.

Fisheries for shell fish are largely confined to narrow coastal zones. Crab and lobster usually inhabit inshore rocky bottom, shrimps and scallops live on sandy inshore grounds, while mussel and oyster are generally associated with sheltered, shallow water such as is found in estuaries. Prawns, however, are found more off-shore on the shelves.

The depth in which some stocks are found varies seasonally. It has been found in the Barents Sea that the catches of cod are better in shallow water in winter, but in deeper water in summer.[38] Seasonal inshore migrations of pelagic species usually in summer result in their being caught in shallow waters during these migrations: this includes varieties like mackerel, sardine, tuna and sprat. Pelagic stocks are often found at different depths in a daily cycle, and the swimming of herring up to the surface at night during migrations is well known.

Depth and bottom conditions also influence the fisheries through the gear which can be employed. Traditional, unmechanized hauling of nets and traps is limited to depths of a few fathoms around the coasts, while lines are rarely hauled by hand in depths of over 50 fathoms. Power-hauling has extended the depths usually worked with lines, shell-fish traps and (most important) trawls. The beam-trawl in the early nineteenth century had a limit of 50 fathoms,[39] but the installation of steam winches at the end of the century allowed most of the shelf area (i.e. down to 100 fathoms) to be trawled; and big modern trawlers often work down to depths of 200 fathoms on the continental slope, while the limit has now been extended to 500 fathoms. Trawlers have also worked increasingly rough bottom with the help of big steel balls on the footrope, although rough rocky bottom remains unworkable; this, however, has its advantages, as the young of some important species (e.g. the cod) seem to be particularly associated with such conditions. Ground-seining is also much influenced by bottom and depth. It is limited to soft sand and mud bottom only, and the ropes with which the net is hauled can be used in 100 fathoms at most. The shallow, sandy bottoms around Denmark were well suited to the pioneering of the modern form of seining.

(d) *Weather & Climatic Conditions*

While these may influence fisheries in a variety of ways, including the fog that hampers visibility, and the effects of heat and cold, the main factor here is the strength of wind and its persistence, in determining both the amplitude of waves and the movement of water masses.

Fog now restricts fishing operations less, as the bigger vessels are now all equipped with radar which minimizes the danger of collision or of running aground. Even so, the additional caution necessary can slow down working, and smaller inshore vessels, for which the extra investment involved in radar equipment is generally uneconomic, are still hampered as before. Most of the coasts of the North Atlantic are subject to fog, and it is especially a hazard in the Grand Banks area, where the juxtaposition of cold and warm water masses, and of contrasting air masses, give the highest incidence of the whole basin. In the North Sea also it can be a considerable hazard, and here the frequency of occurrence is over 20 per cent. in summer in the northern part of the basin, and in winter in the southern part, while the annual frequency over most of the area is between 10 per cent. and 15 per cent.[40]

The influence of temperature is both through fish spoilage in warmer conditions, and of the icing of the superstructures of vessels at the colder margins; to this may be added the restrictions of drift and pack ice, which also hamper operations as far south as Newfoundland in the west, but are no problem south of Jan Mayen in the north-east. The fisheries of the warm temperate eastern Atlantic, where vessels may operate at considerable distances from their bases, are in the greatest danger of having their products deteriorate, and here freezing at sea is usually necessary. Short range operation is still characteristic of the Mediterranean, but here the fish must be processed or consumed without delay. Spoilage in higher temperatures is a danger even in long-range vessels operating in the Arctic, and the extra cost of icing or freezing at sea must be incurred.

Wind frequency and strength play a part in the location of fisheries on both local and regional scales; and there is some seasonal variation in their effects, which over most of the North Atlantic and in the Mediterranean are most marked in winter. The clearest indication of the effects of wind are given by the frequency of gales. Much of the area of the most productive boreal water zone has a particularly high frequency of gales sufficiently strong to interfere with the operation of even the biggest fishing vessels. In the month of February an area stretching from the Grand Banks 3,000 miles to the Barents Sea has a 30 per cent. frequency of gales of force 7 and over (i.e. over 40 knots) which prevents working on open decks. Within this belt in the same month an area stretching some 1,000 miles from south-west Iceland has a 40 per cent. incidence of such gales.[41] Since 1965, records have been published in the West German Fisheries Returns (*Jahresbericht über die Deutsche Fischwirtschaft*) for the numbers of hours of gales of storm force (i.e. wind force 8 and over) at different stations in the North Atlantic. These show totals of around 300 hours per year in the North Sea and North Norway, 500 hours at Faroe, 600 hours on the Flemish Cap Bank, and 1,000 hours in south-west Iceland. The latter figure amounts to 11 per cent. over the year, and in the worst months (January to March) 30 per cent.

or even 50 per cent. of the time may be storm-bound. Even in the northern North Sea, there is a 20 per cent. frequency of storms in winter.[42] Although the most modern vessels—the stern trawlers and shelter-deck liners—do not have crews working on open decks, they may have to stop working because of the possibility of damage to their gear, or even of being swamped. Vessels employing the purse-seine—the most productive modern technique in pelagic fishing—require relatively calm conditions to avoid breaking their gear, and very seldom work when the wind is above force 5 (over 25 knots). Smaller craft, such as the cutters, seiners, liners and drifters used around the North Sea, which usually range from 50 to 80 feet in length, also find this wind strength the general limit to working conditions.

Smaller craft, although they operate closer to land, are more influenced by wind, and round many of the coasts of the North Atlantic they may have their working curtailed at any time of year. In the Mediterranean, the main causes of interruption are in winter when depression tracks traverse the basin most frequently; and the north winds—the Mistral, Bora and Etesian—at this season are particularly troublesome. The Scottish seine-net fleet, the great part of which consists of vessels of 60 to 80 feet, loses at least 30 per cent. of the possible working time in winter; and over the year nearly half the variance of the section of this fleet which is based in the Shetland Isles appears due to wind.[43] In north Norway, the Lofoten cod fishery, which is still mainly prosecuted by small craft of less than 40 feet in an area of frequent gales, always has a number of days when no working is possible. At least 10 per cent. of the working time is usually lost, and at the most exposed outlying stations it is often above 30 per cent.[44]

There is evidence to suggest that wind can also affect the fisheries in less direct ways. Stormy weather after a stock of fish has spawned means high egg mortality during the incubation period, and is a cause of poorer brood strengths. In the East Anglian herring fishery, the catches are usually better in a period of south-west winds, because of the compensating return undercurrent which is set up and brings the herring closer to the shore.

THE INFLUENCE OF LIFE CYCLES AND MIGRATION OF STOCKS ON THEIR EXPLOITATION

Within the life cycles of the different species, seasonal changes nearly always play a significant part. Feeding is usually at its highest level in the warmer part of the year, and fish caught at this season generally have a greater weight; and the season of spawning is also important, as after spawning the spent fish—especially the pelagics—are in poor condition, and deteriorate more easily. It has been estimated that in herring, in the time immediately prior to spawning, the gonads account for 20 per cent. of the total weight.[45] Many of the shell-fish have a moulting season when they cast their shells and seek a refuge which is often inaccessible to the fisherman. Another essential element of seasonal change which has far-reaching consequences is the more or less regular migrations of fish stocks, generally for feeding or spawning: the location of many of the most important stocks changes wholly or in part, and this is especially marked in the pelagics, but also affects demersal species like the cod and plaice, and (more locally) shell-fish, including lobster and crab. These migrations are mainly in response to changing hydrographic conditions and to seasonal availabilities of food. Some of the migrations of the different commercial species are actually inter-related; for example, the mature cod or haddock may follow and feed on shoals of herring or capelin. The longest range regular migrations of commercial fish caught in Europe are reputed to be those of the salmon and the eel. Cases are claimed of salmon crossing from spawning locations in the rivers of western Europe to feeding grounds off south Greenland, and in a classical piece of oceanographic research it was inferred nearly half a century ago that eels migrated from rivers all around Europe to spawning grounds in the Sargasso Sea.[46] Both of these cases, however, are now regarded as controversial.[47]

The best example of seasonal migrations of commercial importance in European fisheries are those of the different herring stocks.[48] In the North Sea stocks, the fisheries have always been concentrated in the warmer half of the year. There are three main stocks: the northern summer-autumn spawners, the central (or Bank) autumn spawners and the southern (or Downs) winter spawners; and although there is some mixing of these, there are discernible migrations of each. In the northern North Sea, the wintering grounds appear to be along the outer edge of the Norwegian Deep, and from here from spring to autumn there is movement in an anticlockwise circuit by the northern stock to spawning grounds off Shetland and the east coast of Scotland and back. They begin this migration when the outflow of a cold nutrient-rich water-mass from the Baltic is set in motion by the spring snow-melt around it: this produces a flowering of phyto-plankton, and in turn more food for the herring is made available. The herring then move westwards, also feeding on sand-eels, and become available to the fishermen in surface waters by April. At this stage they are active and mobile, and the gears required for them are drift-nets and purse-seines. The feeding period is followed by spawning from May to September in locations near the east coast of Britain and there is a subsequent return towards the edge of the Norwegian Deep. They are slower moving at this season, and can be taken by bottom- and mid-water trawls. They appear to winter on the edge of the Norwegian Deep, and although they feed little at this season, the existence here of a long-lived krill species which survives the winter helps to define their location. The other North Sea stocks have similar alternating migrations: the Bank herring spawn on the Dogger Bank in autumn move to wintering grounds in the Skagerrak, and this is followed by a westward feeding migration in spring before returning to spawn on the Dogger Bank. The Downs herring move south to spawn in autumn and winter in the eastern Channel and Southern Bight and are the basis of what is probably the oldest open-sea herring fishery of East Anglia and the Low Countries. The funnelling of the North Sea here helps to concentrate the stock in a limited area, and makes for a more productive fishery.[49] Subsequently they move north-east, and there is a spent fishery off the continental coasts. This is succeeded by a feeding migration north-westwards before gathering to spawn again at the end of the succeeding year.

The Atlanto-Scandinavian herring stocks inhabit a considerably greater sea area than do those of the North Sea, ranging from Spitzbergen in the north to south-west Norway, and from the Norwegian coast in the east to the waters around Iceland. In these stocks the main feeding season is again the summer, which is, however later in the year; and the main spawning time are winter and spring. The best known and most important of these stocks is the Norwegian *vintersild* (winter herring) which begin to approach the coast south of about 63°N. in December, and spawn there between February and April. Spawning formerly took place within the sheltered waters of the skerry guard, and a variety of gears were used including beach-seines and set-nets as well as drift-nets and purse-seines; the latter, operated with the power block has now become supreme, especially as the herring since the late '50s have spawned in the open sea. A distinct stock, the Murman spring herring, spawns off north Norway from March—April in Lofoten to May–August to the east of North Cape. After spring spawning the herring stocks leave the Norwegian coast and migrate towards the waters between Spitzbergen and Iceland, their progress north-westwards probably being governed by the spread of 'biological spring' in these waters which rejuvenates the plankton. A group of herring also gather for winter spawning off the south coast of Iceland, and off the west coast of Scotland; and off north and east Iceland there is a stock which spawns in summer and autumn and gives rise to another fishery.

The migrations of other pelagic stocks define the season and location of their main fisheries. Mackerel collect in shoals in the spring months in deeper water to spawn, as has been shown to the south of Ireland and in the Norwegian Deeps. Thereafter they move slowly shorewards between April and June, feeding on zoo-plankton during a protracted spawning period[50] and may be caught in surface gear like drift-nets. Along many of the coasts from Brittany to south-west Norway, they come close inshore in summer, especially in June and July: at this period they feed on small fish like sprat and sand eels and can be caught by hook and line.

In the warmer waters fished by the European nations, the tuna and sardine are the most important fish. Spawning of the tuna occurs in early summer and the locations are known both in the open sea off Iberia and in the western basin of the Mediterranean. Subsequent feeding migrations bring stocks close inshore, and they are caught in big numbers on the Atlantic coasts of Iberia and on the coasts of the Mediterranean; some of those from the Atlantic move into the Mediterranean at this season.[51] Sardine stocks are confined largely to coastal waters, but they migrate on a local scale, usually into the coast in summer on a feeding migration after spawning. Off the Yugoslavian coast, for example, shoals spawn out in the Adriatic in depths of 30 to 60 fathoms, but move into the coast, where the plankton biomass is highest in summer, to feed.[52]

While less variable in their location than pelagic stocks, there are seasonal movements among such important demersal species as cod and plaice, which can be of vital importance in fisheries. The best known migration of cod is probably that of the Arctic cod in the north-east Atlantic: here there is a spawning migration at the start of the year from the Barents Sea to the coasts of north Norway—especially into the Vestfjorden between the Lofoten Islands and the mainland, and the cod have been fished here commercially for nearly a thousand years. The precise migration route, and the depth and location of spawning in a particular season are largely governed by the junction of cold and warmer water masses, where a temperature of 5°C. to 7°C. obtains.[53] After spawning, part of this cod stock during the return migration comes into the Finnmark coast to feed on the capelin which migrates inshore; this is the basis of the Finnmark spring cod fishery. Later in May to July, trawlers fish a section of the same stock on the Bear Island grounds at the edge of the Spitzbergen Shelf. At this time the cod are concentrated in tongues of warmer water (i.e. with a temperature above 2°C.) thrust up on to the edge of the shelf from the West Spitzbergen current; they are more difficult to catch at an economic rate in the latter part of the year when they have dispersed eastwards with the warming of the waters of the Barents Sea.[54]

In the north-west Atlantic, there is massive inshore migration of cod in summer feeding migrations which follow the capelin, after spawning in spring.[55] This is especially marked in Newfoundland and Labrador, but also occurs in Nova Scotia and Greenland, and thus for two or three months this means that big sections of the stocks are within national fishing limits and outside the reach of European fleets. There are also migrations and spawning concentrations of cod stocks in areas to the south of the main populations in the boreal waters. The spring spawning concentrations in the North Sea on such grounds as the Ling Bank and the Moray Firth are reflected in higher catches at this period, as is the autumn shoaling for feeding. The spring concentration of cod off the western coasts of Iceland for spawning also defines a major fishery effort at that season.

One of the best investigated fish stocks is the North Sea plaice, the life-cycle and migrations of which govern the main location and season of its fishery.[56] The main spawning ground is the Flemish Bight of the southern North Sea, and plaice gather to spawn here from December to February; eggs spawned here are drifted by the prevalent water movement to the shallow nursery grounds on the coasts of Holland and Denmark. After spawning, the plaice themselves resume feeding, and disperse as they do so, most also moving towards the Dutch coast, but others moving north. There is a tendency for them to move away from the coastal nursery grounds as they get older, and an increase in age and size has been demonstrated for the plaice from south-east to north-west in the North Sea. Concentrations thin out towards the north-west, however, and the best plaice fisheries are in the south-east.

Comparable migrations of plaice have been traced to spawning grounds used in winter and early spring in the German Bight, off Flamborough Head, off the east coast of Scotland, and in the Irish Sea; and Icelandic plaice stocks appear to migrate to grounds off the south-west of the island for spawning in spring.[57]

Sufficient is known about other commercial fish to show that the life-cycles of species generally influence the fisheries. Thus, in the North Sea, catches of the haddock show similar spring and autumn peaks to the cod, related to its shoaling for spawning and feeding. With hake, a shoreward movement in spring and summer alternates with an off-shore movement in autumn.[58] The drift migrations also of larvae and young fish from spawning to feeding grounds, instanced above in the North Sea plaice, are familiar in other parts of the ocean, and have been shown in the boreal waters of the Greenland,

Norwegian and Barents Seas for the young of cod, herring, haddock, saithe and rosefish.[59]

As well as the migrations in the open sea, there are migrations of several stocks, which are adapted to warmer conditions, from the Black Sea to the Mediterranean in autumn and back in spring, and these include important commercial species like mackerel and tuna.[60] Migrations from the waters of the Caspian Sea and Sea of Azov of certain stocks, the commercial importance of which is on the scale of some of the main species of the open ocean, are characteristic. These include such as the sturgeon, carp and vobla, and their life cycles have been seriously interfered with by the building of hydro-electric and irrigation installations on rivers like the Volga and Dneiper. As a result catches have greatly decreased and the Caspian Sea now produces annually only about one-half the tonnage which was landed before the First World War. Current plans in the U.S.S.R. aim to enhance the levels of the stocks by a large-scale programme of rearing in hatcheries.[61]

LONG TERM FLUCTUATIONS IN FISHERIES

Fluctuations in the yield of fisheries from year to year, and over longer periods, have always been experienced; and in general these are greater complications for the organization of the industry than are seasonal variations. These long-term fluctuations appear to arise from a combination of hydrographic and biological causes. Changes in the movement and areal distribution of water masses may influence the ecosystems of particular sea areas, and the direction and distance of fish movements. The main cause of fluctuations in the stocks of any area appears to be year-to-year variation in brood strengths. In the inter-war years in the North Sea, for example, cod brood strengths varied by a factor of 12, and haddock by one of 60.[62] The causes of these variations are still generally obscure, and very little is known in detail of the mortality rates between the stages of egg and young adult entering an exploitable stock for the first time. Long-term fluctuations are experienced most in pelagic fisheries, but occur to some extent in all.

In all probability, it is conditions during the early weeks or

E

months of a brood which determine its success: female fish generally lay thousands (or even millions) of eggs, but never do more than a few develop to the adult stage and losses are particularly heavy during the critical infant stages. Stormy conditions, or being blown on to a lee shore, can thin out the numbers at the initial stage; and variations in the available food supply during the larval stage among the plankton is another cause of brood fluctuations. Where spawning grounds are located near the edge of the continental shelf, there is also the danger that a period of off-shore winds could disperse the eggs and larvae into the deep ocean where their chances of survival would be much less.

The herring fisheries are probably subject to the greatest fluctuations. While the nature of the big decrease in the stocks fished in the Sound, which is associated with the decline of the Hansa in the fifteenth century, is obscure, more is known of the more recent variations. At the Norwegian and west Swedish coasts in particular, records are sufficient to show alternating periods of scarcity and abundance over several centuries: in Norway periods of 50–80 years of good fisheries have alternated with lean periods of 30–60 years.[63] Although the periods of success in west Sweden have corresponded with those of Norwegian failures and vice versa, this cannot be because of the same stock changing its position, as the herring caught at the Norwegian coast are an Atlanto-Scandinavian stock, separate from the North Sea stock caught in Sweden. This rather suggests that long term hydrographic changes in the whole area give conditions which simultaneously are favourable to the one stock and detrimental to the other.

The above fluctuations are related to changing migration patterns of the stocks, but the influence of varying brood strengths is well attested also in the herring fisheries. A very prominent brood in the Norwegian winter fishery in the early years of this century was that of 1904, which dominated the year classes from 1908 to 1918 and could still be distinguished up to 1924.[64] On a lesser scale the 1928 and 1949 broods could be recognized over series of years in the Buchan fishery of Scotland,[65] and good broods occurred at almost regular three-year intervals in the East Anglian fishery from 1921 to 1932.[66]

Variations in other pelagic fisheries are also well known.

The pilchard fisheries of Cornwall and southern Ireland have a history of fluctuations; this is partly due to these waters being near the northern limit of the range of the species. The only period of abundant shoals on the southern coast of Ireland this century were the years 1935 to 1939, while in Cornish waters a period of big numbers in the early years of the century was followed by a run of lean years until around 1950.[67] The related sardine stocks off the French coast have also shown periodic failures, such as that which occurred in the early 1900s.[68]

While variations in brood strengths are less important among demersal fish, and have received rather less attention from fisheries biologists, they are known, for example, among the haddock in the different parts of the north-east Atlantic. Here the years of good broods have been shown to differ for separate stocks, showing the importance of local conditions. The best brood of recent years in the North Sea was that of 1962, which showed itself prominently in the catches of the latter 1960s.

Among the white fish stocks, the most important changes in location of fisheries have taken place on the northern margins of the exploited area in the Arctic, and appears to be caused by variations in the flow of water-masses. The cod fishery at Bear Island failed about 1900, but was resumed after 1925 when stocks returned at the same time as a more powerful movement of the North Atlantic Drift was noted.[69] At the same time the increased circulation of the North Atlantic is probably responsible for bringing cod stocks to West Greenland, and both these fisheries have been sustained till the present time. Within the wider regional boundaries of the cod fisheries, the location of the fishery of a particular season may be influenced by water-mass movements. Thus in the north Norwegian winter fishery, a strong northward movement of the coastal current forces the southward moving cod off-shore and results in poorer catches for the inshore fleets. On the spawning grounds themselves the thermocline which governs the fish depth (page 51) may vary from 20 to 80 fathoms below the surface. An increase of cod stocks in the Baltic Sea has also been observed in the present century, and in the last 20 years they have entered the Gulf of Bothnia for the first time.[70]

REFERENCES

1   R. Morgan, *World Sea Fisheries*, 1956, p. 22.

2   A. Hardy, *The Open Sea*, vol. I, 1956, p. 52.

3   T. Laevastu, 'Natural Bases of Fisheries in the Atlantic Ocean', in *Atlantic Ocean Fisheries*, ed. G. Borgstrom and A. J. Heighway, 1961, p. 21.

4   L. Zenkevich, 'Biological Appraisal of the Ocean, and the Problem of Transoceanic Acclimatisation', in *P.P.I.T.C.C.L.R.S.* (Papers Presented at the International Technical Conference on the Conservation of the Living Resources of the Sea), 1956, p. 139.

5   L. Zenkevich, *Ibid.*, pp. 133, 139.

6   H. Tamb-lyche, 'An Outline of Marine Biological Sciences', in *Some Aspects of Fisheries Economics*, III, ed. G. M. Gerhardsen, 1964, p. 123.

7   T. Laevastu, *Ibid.*, p. 18.

8   B. Rasmussen, 'Marine Research as Applied to Fisheries in Norway and in the Atlantic Area in General', in *Some Aspects of Fisheries Economics*, III, ed. G. M. Gerhardsen, 1964, p. 136.

9   D. H. Cushing, *The Arctic Cod*, 1966, p. 66.

10   O. A. Popova, 'Some Data on the Feeding of Cod in the Newfoundland Area of the North-west Atlantic', in *Soviet Fisheries Investigations in the North-west Atlantic*, ed. Y. Y. Marti.   English Edition 1963, p. 228.

11   W. C. Hodgson, *The Herring and its Fishery*, 1957, p. 151.

12   B. B. Parrish, 'The Cod, Haddock and Hake', in *Sea Fisheries. Their Investigation in the United Kingdom*, ed. M. Graham, 1956, pp. 312–33.

13   H. Wood, 'Fisheries in the United Kingdom', in *Sea Fisheries*, ed. M. Graham, 1956, pp. 43, 44.

14   D. H. Cushing, *Ibid.*, p. 43.

15   B. B. Parrish, *Ibid.*, p. 269.

16   O. A. Popova, *Ibid.*, p. 236.

17   B. B. Parrish, *Ibid.*, p. 270.

18   For maps showing distribution of cod, haddock, herring and rosefish in the North Atlantic, see Y. Y. Marti, 'Some Similarities and Differences in Conditions under which Boreal Fish Species exist in North-east and North-west Atlantic', in *Soviet Fisheries Investigations in the North-West Atlantic*, ed. Y. Y. Marti.   English Edition, 1963, pp. 59–64.

19   J. Furnestin, 'The Atlantic Sardine,' in *P.P.I.T.C.C.L.R.S.*, 1956, p. 92.

20   W. F. Royce, 'A Statement on the Ecology of Tunas', in *P.P.I.T.C.C.L.R.S.*, 1956, p. 118.

21   B. B. Parrish, *Ibid.*, p. 271.

22   A. J. C. Jensen, 'Reactions of Mackerel to Environmental Factors', in *P.P.I.T.C.C.L.R.S.*, 1956, pp. 97–98.

23   F. R. Harden-Jones, *Fish Migration*, 1968, p. 146.

24   O. A. Popova, *Ibid.*, p. 229.

25   R. Kändler, 'Response of the Plaice to Environmental Factors', in *P.P.I.T.C.C.L.R.S.*, 1956, p. 105.

26   R. Kändler, *Ibid.*, p. 105.

27 O. A. Popova, *Ibid.*, p. 229.

28 B. Rasmussen, *Ibid.*, p. 137.

29 B. B. Parrish, *Ibid.*, p. 270.

30 A. J. C. Jensen, *Ibid.*, p. 98.

31 R. Kändler, *Ibid.*, p. 102.

32 M. M. Ovchynnyk, 'Development of Some Marine and Inland Russian Fisheries and Fish Utilisation', in *Atlantic Ocean Fisheries*, ed. G. Borgstrom and A. J. Heighway, 1961, p. 272.

33 W. C. Hodgson, *Ibid.*, p. 98.

34 Y. Y. Marti, *Ibid.*, p. 62.

35 A. Hardy, *Ibid.*, p. 86.

36 R. Morgan, *Ibid.*, p. 208.

37 M. Graham, 'Plaice', in *Sea Fisheries. Their Investigation in the United Kingdom*, ed. M. Graham, 1956, p. 333.

38 F. R. Harden-Jones, *Ibid.*, p. 149.

39 C. L. Cutting, *Fish Saving*, 1955, p. 233.

40 T. Laevastu, *Synopsis of Information on the Oceanography of the North Sea.* (F.A.O. Fisheries Division, 1960). Table 3.1.

41 T. Laevastu, *Natural Bases of Fisheries in the Atlantic Ocean*, p. 28.

42 T. Laevastu, *Synopsis of Information on the Oceanography of the North Sea.* Table 3.1.

43 C. A. Goodlad, *The Geography of Shetland Fisheries* (Ph.D. thesis University of Aberdeen, 1968, pp. 358, 359.

44 *Lofotfisket*, annual reports.

45 F. R. Harden-Jones, *Ibid.*, p. 89.

46 J. Schmidt, 'The Breeding Place of the Eel', *Phil. Trans. Royal Soc.*, *B.* 211, 1922, pp. 179–208.

47 F. R. Harden-Jones, *Ibid.*, pp. 63–64, 69–85.

48 W. C. Hodgson, *Ibid.*, pp. 16–25.

49 W. C. Hodgson, *Ibid.*, p. 149.

50 A. Hardy, *Ibid.*, pp. 78–80.

51 A. Hardy, *Ibid.*, p. 81.

52 T. Gamulin, 'Migrations of the Adriatic Sardine in Relation to Zooplankton,' in *P.P.I.T.C.C.L.R.S.*, p. 338.

53 G. Rollefsen, 'The Arctic Cod', in *P.P.I.T.C.C.L.R.S.*, p. 118.

54 D. H. Cushing, *Ibid.*, pp. 40–43.

55 F. R. Harden-Jones, *Ibid.*, pp. 156–58.

56 M. Graham, *Ibid.*, pp. 337–40.

57 F. R. Harden-Jones, *Ibid.*, pp. 168–79.

58 B. B. Parrish, *Ibid.*, pp. 302, 303.

59 Y. Y. Marti, *Ibid.*, p. 65.

60 M. M. Ovchynnyk, *Ibid.*, p. 295.

61 M. M. Ovchynnyk, *Ibid.*, pp. 269, 280, 281.

62 B. B. Parrish, *Ibid.*, p. 323.

63   A. C. Simpson, 'The Pelagic Phase', in *Sea Fisheries. Their Investigation in the United Kingdom*, ed. M. Graham, 1956, p. 236.

64   A. Hardy, *Ibid.*, pp. 58–60.

65   B. B. Parrish and R. E. Craig, 'Recent Changes in the North Sea Herring Fisheries', *I.C.E.S. Rapp. et Proc-Verb.* vol. 143, 1957, p. 16.

66   W. C. Hodgson, *Ibid.*, p. 95.

67   A .Hardy, *Ibid.*, p. 86.

68   A. C. Simpson, *Ibid.*, p. 237.

69   A. C. Simpson, *Ibid.*, p. 236.

70   W. R. Mead, *An Economic Geography of the Scandinavian States and Finland*, 1958, p. 178.

CHAPTER IV

The Historical Development of the Fisheries of Europe

While the present patterns of resource exploitation, transport and consumption in the fisheries of Europe are in large measure the creation of the greatly accelerating pace of development during the twentieth century, they have been built on foundations laid over a long historical period, and considerable sectors of the patterns are the legacy from previous centuries.

Fisheries had a place throughout the continent in the earlier subsistence economies, and the fisheries prosecuted, boat and gear types employed and methods of fish preservation all had links with environmental conditions in different regions. Commercial fisheries also have a history of many centuries, and settlements of specialist fishers are reflected in place-names such as Piscaya in Italy and Fisherrow in Scotland, while Icaria in Greece indicates a coast rich in fish. Fishery resources have been seen by some as a main factor in the course of history, and it is certainly the case that the strengths of the various sea-powers of Europe have been functionally linked to the development of their fisheries. There is thus ample evidence that fisheries played an important—and occasionally a key—role in many communities from the earliest times, although any absolute measure of the importance of fisheries in Europe have to wait till recent times, with the rise of more systematic and comprehensive recording of data during the industrial era.

FISHING IN THE PREHISTORIC PERIOD

Archaeological evidence shows that fish have been exploited in Europe for thousands of years, and they were part of the diet of most of the hunting groups who populated the continent before the coming of farming.[1] Shell-fish have been found associated with remains of Neanderthal man in caves at Gibraltar, and in the more elaborate cultures of groups of 'homo sapiens' in the Upper Palaeolithic—such as the Magdalenians in France and Spain—have been found remains of vertebrate fish. These consisted mainly of the more easily exploited species from inland waters, especially pike; but sea fish, among which wrasses are prominent, are also known, some of them occurring at inland sites, showing transport or possibly even trade in fish. The Palaeolithic shades into the Mesolithic with the close of the Ice Age, and a characteristic of Northern Europe at this stage was the series of coastal communities, from Brittany to the Baltic, who subsisted mainly on fish. The land flora and fauna were still relatively poor in the north at this period, and groups such as the Ertebölle in Denmark and the Larnian on the shores of the North Channel (between Scotland and Ireland) lived largely on shell-fish, although at least some seem to have ventured out in boats to catch free-swimming species such as the cod in deeper water. It appears that most basic fishing techniques, including spears, gorges, hook and line, traps and nets, as well as boats, were in use before the spread of farming, although boats particularly were to be more fully utilized later.

While the relative importance of fishing declined with the development of farming cultures, it continued to play a significant part in the subsistence economies of Europe in both prehistoric and later times. At the subsistence level fisheries were prosecuted by communities who also practised farming (i.e. crofter-fishers). This type of joint-occupation was a very obvious way of achieving greater variety in diet, and indeed most fisheries were prosecuted in conjunction with farming right up until the Industrial Revolution, even when trade in fish was on a considerable scale. In most parts of the continent, however, fishing has been regarded as an occupation of the lowest social classes.

FISHERIES OF THE CLASSICAL PERIOD

It is, of course, in Mediterranean Europe that the first documentary records of fishing occur.[2] Even in times when the natural vegetation and soil cover of the Mediterranean lands was more complete, with a resultant higher carrying capacity for game and livestock, fisheries were always of considerable importance from Mesolithic times onwards. The increase in population and the spread of city civilization must have generated pressure on available sources of flesh foods, and Classical Times certainly appear to have witnessed an expansion of effort in fisheries. Both fresh and preserved fish were dietary items, and it is a comment on the problems of transport that fresh fish were generally luxury food of the wealthy, while preserved varieties were eaten by the poorer classes and slaves.

While there are records which show something of the scale and organization of the fisheries (which were largely in rivers) of the civilizations of Mesopotamia and Egypt, the first records of fisheries prosecuted on a large scale in Europe are from Phoenician sources; the Phoenicians had bases in Spain from which they operated, both within the Mediterranean and in the open Atlantic, and there was an organized trade in preserved fish, the destinations of which included their cities in the eastern basin of the Mediterranean. While Homeric sources in Greece show a distaste for fish and suggest a low level of activity in fisheries, by the sixth century B.C. catching and trade in fish were prominent in the economy of Greece and of her overseas colonies.[3] The rich and varied fisheries of the Sea of Azov were a main economic incentive to Greek colonization in Scythia, and their commercial development in the Ancient World has been likened to the opening up of the Grand Banks fisheries of Newfoundland by the West European powers during and after the Great Age of Geographical Discovery. The trade in fish between Scythia and Greece continued unbroken until the Turks got control of the passage between the Mediterranean and Black Seas in the fifteenth century. The rise of Rome, with her extensive and unified empire centred on the Mediterranean, led to a greater scale of exploitation and trade in fish. Preserved fish were brought to Italy from all parts of the Basin, and Rome herself with her big slave population was an almost insatiable market.

In the Mediterranean Basin the bulk of the fisheries were inshore; tuna was the most valuable and sought-after species, being largely caught in coastal traps and nets as it swam near the coast. It was most caught in the western basin where it is more abundant, but also in the open Atlantic beyond Gibraltar. A variety of other fish including red mullet, hake, sardine, mackerel, eel and sword-fish were also caught. Shell fish (at the coast and in estuaries) and river fish added to the resources, and here some management was practised, especially by the Romans, which included oyster-farming and restocking of inland lakes and streams. In Italy thousands of slaves were trained as fishermen, and many of them worked in enterprises which were organized on a big scale. While there was undoubtedly a considerable consumption of fresh fish in the vicinity of catching points, the great part of fish which entered into recorded commerce was preserved, and for this purpose a great deal of salt was employed, most of it being made by evaporation in coastal salt-pans. Fish were preserved by splitting, salting and drying and—for the more fatty species like mackerel—by pickling in brine in earthenware jars or barrels; smoking was also used to a lesser extent. The luxury item of fresh fish had to be brought to the market in special well-boats, and the price appears to have been several times that of preserved. The setting aside of certain days in the Roman calendar when salt fish were the only flesh food permitted was linked to economic necessity, and anticipated the rules of the Church in regard to meatless days and Lent in later centuries.

GROWTH OF FISHERIES
BEFORE THE INDUSTRIAL REVOLUTION

Despite the greater wealth of marine life on the continental shelves bordering the Atlantic Ocean, difficulty of exploitation and remoteness from main population centres meant that the fisheries of the Mediterranean were the chief ones of the European continent till at least the early Medieval period. Even so, the sparse records of the Dark Age period suggest that by the sixth or seventh century the fisheries on some parts of the Atlantic coasts of the continent were of considerable importance, although river and lake fisheries almost certainly were the main ones exploited at this time, and indeed continued to be so in much of eastern and central Europe till modern times. In the sixth century there is evidence for some trade in fish on the Atlantic coastlands, in which the English, French and Flemish figured;[4] and in the seventh century on the Friesian coast there were communities whose main activity seems to have been in fishing. These early fisheries were probably all inshore, and the Vikings and the Normans are credited with the first ventures in deep-water fisheries: in the tenth century the Normans were pursuing deep-water fisheries, some of them probably with long-lines, from seasonal bases in the south of Ireland, and in the following century English fishermen were venturing into deeper waters for herring.[5] The early commercial fisheries from the west coasts of Europe were principally for cod and herring, which are still the main species sought by the modern fleets; these are of such major importance in the development of the continent's fisheries that they are discussed separately (pp. 71–81).

The rise of commercial fisheries, like so many economic activities, is closely bound up with the rise of towns and cities. There are many examples of coastal towns which had their own communities of fisherfolk, although rural crofter-fishers were also known throughout the continent. Enterprise in trade was largely concentrated in towns, especially through gild organizations, which were especially important in main centres like Santander, Yarmouth and Lübeck, and also at a great inland market like Cologne; and London had its own fishmongers gild as early as 1154.[6] As well as dealing in fish, town centres also played a key role in the distribution of such essentials as salt for preservation, staves and hoops for barrels, linen and hemp for lines and nets and iron for hooks; and they were often dominant in the building of boats. Salt was always of central importance in commercial fisheries, being for the most part indispensable for preservation; and it figured considerably in political controversies, as taxes were often imposed on it as a source of revenue.

While the early history of fishing craft is little known, there are boat-types in the Mediterranean which are descendants of those of Classical times, while in north west Europe the Vikings, with their strongly-built double-ended craft are credited with having a strong influence in boat design. Despite the early deep-sea ventures of fishermen from Normandy, the main innovators here were the Basques and the Dutch. The former

were by the twelfth century exploiting whales and cod in the
Bay of Biscay, and subsequently extended their fisheries
throughout the north-east Atlantic as far as Norway.[7] The
Dutch evolved a superior class of vessel (the 'buss') for fishing
herring with drift-nets, and were the first to use long lines on a
big scale in white fisheries. While the dates of the beginning of
such advances cannot be ascertained with any precision, their
herring fisheries were certainly growing from the fourteenth
century, and their prowess with long lines was proved by the
seventeenth century.[8] Fishing craft also benefited from the
innovations made during the Age of Discovery, and subsequently fishermen in distant-water fisheries were equipped with
more seaworthy vessels, capable of tacking into the wind, and
were better able to navigate.

While fisheries on Icelandic grounds seem to have been
known from the twelfth century, it appears to have been in the
fifteenth century that they were first exploited with any
intensity by fishermen like the Basques, English and French;
and with the rapid development of the Grand Banks in the next
century, all the major grounds at present worked by the
European fishing nations, apart from those of the Arctic and
South Atlantic, were known.

Of other sources of fish to the European nations, the Caspian
was for centuries the location of the main fisheries of Russia, and
was a source of sturgeon, herring and other fish. Big-scale
fisheries are on record at Astrakhan in the mid-sixteenth
century, and the Caspian continued to provide the main
supplies for Russia right until the inter-war years of this century;
in 1913 it was estimated as producing 65 per cent. of the total.[9]
Rivers and lakes have always been important sources of fish,
especially in Russia and the other parts of eastern Europe, and
this was the case in the Slav-inhabited areas even before the
establishment of towns in the Medieval period. In the big
variety of river fish, salmon, trout and sturgeon have probably
been most important, and salmon fisheries especially were often
reserved as of right for feudal kings and lords. Even in a sea-
surrounded country like Scotland, salmon were the subject of
the main Medieval commercial fisheries, and the operation of
them is still governed by statute. It is a comment on changing
values that salmon, formerly food for the poorer classes which

aroused numerous complaints by its frequency in their diet in a pickled form, became in modern times a luxury product for the wealthy when it would be transported direct to market in ice or frozen.

FISHERIES DEVELOPMENT IN THE INDUSTRIAL AGE

The modern phases in the development of European fisheries show the acceleration of the tempo of economic activity that is prominent in so many fields: they have witnessed a more intensive exploitation of fishery resources, together with a generally increased range of fishing craft from their bases. Associated with this have been outstanding technical innovations together with developments in marketing and commercial organization. The locations of the bases from which these modern fisheries have been prosecuted have in broad terms remained those of the preceding centuries, although there have been changes in relative importance as between countries and between ports. There has been one really major change here in the recent growth of big fisheries bases on the European coast of Russia, as part of national policy in expanding protein food supplies. In fisheries, modern times have naturally witnessed great improvements in catching, especially with the installation of engines in boats, the power hauling of gear and the development of many scientific fishing aids. These have been functionally related to improvements in the shore side of the industry, including better processing methods, more sophisticated engineering and maintenance facilities, and—of over-riding importance—more rapid transport to markets, especially those in towns and cities where there is a concentration of consumer demand at particular points. These modern changes have been prominent in both demersal and pelagic fisheries, and (in the most recent phases) show prominently in the very valuable product of shell-fish.

In the earlier phases of the modern growth of fisheries, developments in the herring fisheries in the countries around the North Sea were of primary importance. Here the locations of exploitation remained substantially as before, and techniques of catching improved only in limited degree: more important were developments in commercial organization, including the strengthening and multiplication of links within Europe, and the forging of contacts with regions involved in tropical plantation economies, especially the West Indies.

In white fisheries, modern growth centres on the adoption of trawling as the main catching technique:[10] this has radically changed the scale and pattern of activity by the progressive outwards extension of the sea areas intensively fished. While gear of this type appears to have been used at least from the fourteenth century, (the 'wondryschoun' in England)[11] its use on a significant scale is first recorded on the southern North Sea among Dutch, Belgian and French fishermen in the seventeenth century; but its modern development was substantially English. Trawling in the latter eighteenth century became the main technique at Brixham in Devon, and fishermen from the south-west settled in Kent in the early nineteenth century and employed the trawl to serve the lucrative London market. Although early efforts were with the small and rather clumsy beam trawl, and the smacks involved in it might only manage one drag per day and needed a fair wind to make any speed with the drag of the gear, its superiority to lining was clearly demonstrated: it hunted—rather than trapped—the fish. In the 1830s, the fleet system of operation was introduced from east coast ports, by which fleets of trawlers could work in an area of perhaps 10 miles radius around a fast cutter which collected their catches and took them to market. Off-shore working in fleets, however, inevitably increased the time-interval between catching and consumption, and created the need—especially in summer— for fish to be gutted aboard and for ice to inhibit spoilage. In the nineteenth century the ice was got by collection in winter and by import from Norway, and was kept in insulated stores, but from the end of the century ice was made commercially by artificial means. Icing in fact eliminated a bottle-neck in the links to market, as salting and drying had inevitably been labour-intensive and time consuming. Subsequent to the mid-nineteenth century the proportion of salt fish reaching markets in Britain greatly declined, and in the twentieth century the marketing of white fresh fish developed progressively in other European countries. Another result of trawling and icing was a widening in the range of species delivered to market: the early trawlers were catering for a market accustomed to cod and ling, and rejected as offal about three-quarters of their catches, including such species as haddock, plaice and halibut, which had not been suitable for salting in earlier times; but such species were to become prime fish.

The spread of the railway network, which allowed a much wider market to be reached with iced fish from the coast, greatly stimulated trawling, as did the build-up of industrial city populations, which meant that consumption became concentrated more at particular points. Hull (from the 1840s) and Grimsby (from the 1850s) became the main centres of trawl operation, and their hinterland reached by rail delivery included the industrial areas of the West Riding and the Midlands as well as the London area.[12] It was, however, the application of steam-power to the vessels themselves which really made the superiority of trawling over other methods of white fishing decisive. Steam carriers were used from the 1860s to establish more reliable links with port markets, and from the 1880s steam power was successfully installed in trawlers—with spectacular results. Subsequently the catch of the individual boat was multiplied several times, the operation was much less at the mercy of wind and tide, the range and sea area of operation were increased, the steam winch allowed trawling at greater depths than the former limit of 50 fathoms, the size of vessels increased, and they began to be built in steel. From the 1880s steel warps replaced ropes for hauling the trawl and allowed working down to 200 fathoms; and the beam trawl was replaced by the otter trawl (which was more easily handled and could be made considerably bigger) in the 1890s. In the last twenty years of the century other main bases of the trawl fishery arose rapidly at Aberdeen, Fleetwood, Milford Haven and North Shields, and these allowed the catches from around most of the British coasts to be delivered rapidly to market via the railway network. 'Single-boating' replaced fleet operation, which eliminated awkward trans-shipment in rowing boats, and the individual trawler had greater freedom of movement and could carry an economic catch on its own. Fish auctions were established at all the main trawl ports, and proved the most efficient means of regulating supply, demand and prices. During the nineteenth century, the limits of the area regularly worked by trawlers in the North Sea expanded steadily northwards until by the end of the century grounds on the latitude of the north part of the Scottish mainland were regularly fished, and if west coast grounds were on the whole less intensively trawled, vessels from Fleetwood and Milford Haven were ranging over them widely. By the last decade of the century, too, Hull and Grimsby trawlers were regularly working at Iceland, and in the early years of this century they reached the Barents Sea. There was an inevitable hiatus in expansion during the First World War, but in the inter-war period the main grounds worked by the now supreme trawling ports of Hull and Grimsby became the richer distant-water ones of the Arctic—Iceland, the Norwegian coast and the Barents Sea.

The transition to a more technological and capital-intensive fishery was achieved from the latter nineteenth century also in Germany; here it was aided by a series of measures for political and customs unification taken between 1828 and 1871, and also by the rapid growth of industrial population concentrations along with the extension of the railway system. Germany rapidly reached a position of being largely self-sufficient in fish supplies, having formerly been a big importer. While power-driven craft were adopted to some degree in most countries from the early years of the twentieth century, this largely took the form of the installation of petrol or paraffin engines in the traditional small inshore craft; this was substantially the case even in such important fishing nations as Norway and Portugal, where forces of conservatism allied to shortage of capital greatly restricted the deployment of modern methods till the mid-twentieth century. While in other countries, like France, the Netherlands and the U.S.S.R., some use of steam-powered fishing craft was made from the inter-war years, a significant development of that period was the first installation of diesel power, which has become the main source of motive power in the world's shipping, including its fishing fleets. Another important innovation of this period was the beginning of the big Russian expansion in deep water fisheries in 1929.[13]

The period since the Second World War has witnessed an unparalleled acceleration of development in fisheries, as in other fields of economic activity. It has seen a considerable modernisation of the fisheries of nations traditionally involved, like

Norway, France and Spain, together with the appearance in force of the countries of Eastern Europe, in which big programmes of fisheries expansion have been part of national development plans: the U.S.S.R. now far outstrips all other European nations, while East Germany and Poland have also constructed large modern fishing fleets. Fisheries expansion has figured on a more modest scale in economic growth in southern Europe in Italy, Greece and Yugoslavia.

Increase in efficiency has occurred notably in pelagic fisheries from the inter-war period, when the catching of herring by bottom trawl was pioneered in Germany, and followed by the Dutch and others. Subsequently the mid-water trawl was developed, and in post-war years this has become a main technique. It is now, however, overshadowed by the purse-seine—gear used since last century in the Norwegian fjords, but now—thanks to power-hauling and lighter synthetic fibres for the net—capable of deployment in the open sea on a grand scale. Steel boats have been used in fishing since the end of last century with the building of bigger trawlers, and although the majority of European fishing craft are still of wooden construction, the great part of the catching power is now vested in big steel vessels, and nearly all craft above about 80 feet are steel-built.

In demersal fisheries, there has been an extension of more intensive exploitation into the outlying sea areas of Greenland, and more operation in the Barents Sea and Newfoundland; this has been accompanied by the further development of vessels—especially trawlers—better adapted to distant-water working; stowage space has been enlarged and freezing equipment installed, while processing aboard has included more fillet production and fewer round fish. All this allows vessels to remain several months at sea, and so to minimize the adverse economic effects of operating at long range. In addition, big modern trawlers are stern-operating, which allows more mechanical handling of the gear and allows safer working in adverse weather conditions. The big freezer trawlers are now operated above all by the U.S.S.R.; and the eastern European countries especially have developed fleet operations in both demersal and pelagic fisheries, supported by mother ships and carriers to take catches to market. Machine-gutting is already well-proved on land, and is now beginning to be installed on board,

F

where it cuts the monotonous and labour-intensive work by hand.

An important development in shorter-range demersal fisheries has been the development of seining by smaller craft, originally in Denmark, and adopted in Scotland and elsewhere: on suitable ground it is more productive than trawling, and requires less powerful vessels.

Post-war years have seen a range of other innovations that have increased the productivity of fishing fleets. These include higher-compression and higher-speed engines, which means that the engines take up proportionately less space; hydraulic winches and steering; echosounders and asdic—of great value in showing water depth and in locating fish; radio telephones, radar scanners and precise navigation aids like the Decca and Loran systems. As well as improving efficiency, these innovations have inevitably generated greater competition at both national and international levels in fisheries.

Modern times have witnessed extensive developments in marketing, through the rapid transport made available by rail, road (and on occasion now by air). As a result a much bigger proportion of the catches has reached the consumer fresh, while the traditional preservation techniques of drying, salting and smoking have been supplemented—and partly displaced—by more sophisticated methods including canning and freezing. In large measure coastal auction markets have been established to handle first sales, especially in the bigger and more developed countries, and marine engineering and other facilities needed by the more complex modern fleets have been largely concentrated at the same points.

In Eastern Europe, a bigger proportion of the catches still reach the consumer preserved in more traditional ways, and distribution is still largely by the rail network. While the economic viability of the big fishery enterprises built up here in post-war years cannot be readily assessed, it has been stated that the basic explanation of the big expansion of the U.S.S.R. is that in Russia protein foods can be more cheaply produced from the sea than on land.[14] It appears that with the present level of development in Eastern Europe, fisheries may in fact be more profitable than in the West. Small-scale local catching and marketing has survived most in the Mediterranean area, although Italy has developed some large trawler fisheries off Algeria and Mauretania. A continuing feature of fish supplies in the Mediterranean has been the import of salted cod from Norway, Iceland and the Faroe Islands, and these countries have also made inroads this century with dried cod in markets in West Africa and South America—especially Nigeria and Brazil.

THE HERRING FISHERIES

The herring fisheries were from Medieval times till the start of the twentieth century the main commercial ones in Europe. The main seats of enterprise have always been the coasts of the north-west, which is the part of the continent fronting the big sea area in which the shoals live and migrate. The location of the main land bases of the fishery has changed several times, partly with the rise and fall of particular sea-powers in history, but also with changing abundance and patterns of migration of different herring stocks. In turn herring fisheries and trade were dominated by the Hanseatic towns in the Medieval period, and by the Dutch from the sixteenth century to the early eighteenth century; in the latter half of the eighteenth century the fisheries of the Bohuslän coast of Sweden rose to prominence, to be succeeded in the following century by the Norwegians and British. At the present day all the coastal nations in north-west Europe are engaged in the herring fisheries, and the Russians and Norwegians vie for leadership, although the scale of operation is at present greatly restricted by the stocks being at a low level. Both inshore and open-sea fisheries have been important, with the general emphasis going more with time to the latter, as bigger and more reliable catches could be got by seeking out the variable shoals. The scale of the fisheries and of the accompanying shore organisation has also grown over the centuries, and there has been considerable trade in ancillary materials. The great importance of the fisheries has been due above all to the scarcity of flesh foods in Europe at a simpler—and largely subsistence—economic stage, and was enhanced by the practice of abstaining from meat during Lent and on fast days of the Church: in much of northern Europe, pickled herring has been a staple food-stuff.

While the beginnings of commercial herring fisheries are lost in history, there are records of their prosecution at least from the sixth century in East Anglia, and in the ninth century in Egil's Saga.[15] Their first location on a big scale, however, was in the Sound, where they were the main economic foundation of the Hanseatic towns, and where there is evidence for their vigorous prosecution by the year 1200.[16] Basically this was an inshore fishery operated by Danish fishermen using seines and drift-nets, but financed by capital from Lübeck, Bremen and the other Hanseatic towns, and using Lüneburg salt from North Germany to cure the catch ashore. Little detail is available on Hanseatic curing, but it appears that herring were often ungutted, which must have limited their keeping qualities; and surplus catches were boiled for oil.[17] It was an autumn fishery, and there was generally time to cure the catch and transport it to destinations for consumption over the winter. The area included in the herring trade of the Hanseats included a substantial part of northern, western and central Europe: the main trade lines were up rivers like the Elbe, Weser and Vistula into Germany and Poland, but also eastwards into Russia, and west to Flanders, France, England and even Spain and Portugal. The ultimate decline of the Hanseatic herring trade is bound up with a period of failure of the herring shoals in the early sixteenth century; and although there have been herring fisheries in the Baltic since, the dominant position was permanently lost.

When the Dutch succeeded the Hanseats as the leading herring fishers, there was a change in the location of operations as they operated in summer and autumn in the North Sea with bigger craft, and there was some realignment of the marketing links as their main distribution network was via the Rhine, Meuse, Scheldt and Elbe. Hamburg was a main foreign market, but they had, too, a trading network which included the coastlands of Europe from the Baltic round into the Mediterranean, with a return cargo of 'Lisboa' salt for curing. The dominance of the Dutch over a period of two centuries largely reflected a better technique of exploiting the resource and of preserving the catch: their 'busses' were boats of 80 to 100 tons or more, with crews of 14 and 15[18] and were capable of ranging over the North Sea and of staying at sea for months at a time: indeed their main fisheries were around Shetland. It was the Dutch, too, who developed and perfected the technique of drift-netting in the open sea; this involved the setting of a train of nets, perhaps half a mile long, in the sea which hung vertically downwards from floats on the surface and the fish were caught as they swam up to the surface in the hours of darkness. The name of one Beukels who lived in the fourteenth century is associated with the Dutch technique of packing alternate layers of gutted herring and salt in barrels—a practice which was performed at sea, and busses could carry up to 35 lasts (70 tons). The herring were divided into a series of classes, and there was a system of government inspection to guarantee quality. At its peak in the golden (seventeenth) century of the Dutch, this fishery was on a remarkable scale for its time, and there were up to 2,000[19] busses involved, each of which made three voyages per year. A record from 1614 shows 150,000 tons exported, and in 1669 450,000 people were employed ashore or afloat in it— some 20 per cent. of the Dutch population:[20] the herring fishery was indeed the foundation of the maritime supremacy which Holland enjoyed for a time among the sea-powers of Europe.

The eighteenth century saw a fairly rapid decline of the Dutch as a maritime and fishery power, and the mounting of a challenge to them by the British, Norwegians and Swedes. In the second half of the eighteenth century, the most successful of these were the Swedes, who capitalized on the big shoals that frequented the Bohuslän coastal archipelago in autumn and winter.[21] This was an inshore fishery operated mainly with beach seines, and the work of curing and boiling for oil was done ashore. It has been shown that during the three-month season workers frequently earned as much as three to five times as much as Swedish agricultural labourers made in a year. In these circumstances there was no difficulty in recruiting a labour force, which totalled about 16,000 (including 5,000 fishermen) at its peak, and which consisted mainly of seasonal migrants. Up to 120,000 barrels were produced yearly: at this period Sweden dominated the Baltic herring market, and sent oil to Germany and various parts of Western Europe.

When the shoals deserted the Bohuslän coast in 1808, it was the turn of Scotland and Norway to dominate the herring fisheries: in the same year of 1808 winter herring returned to the coast of south-west Norway after a lapse, while in the following year a new bounty system in Britain greatly accelerated the tempo of activity. By the 1860s, 82,000 fishermen in Norway were seasonally employed in an inshore fishery which employed seines, set- and drift-nets, and some 600,000 barrels a year were exported, in addition to a great deal consumed in Norway.[22] Britain, and more particularly Scotland, also rose in the nineteenth century, [23] and was particularly prominent after 1870, when the big shoals forsook the Norwegian coast for the remainder of the century. Herring had been one of Scotland's main commercial products even in Medieval times, although the scale of operations was limited. The nineteenth century at last saw the mounting of an organization capable of a big expansion, and after the priming of the pump with bounties, these were removed in the 1830s, but a self-sustaining growth continued until the First World War. As with the Dutch, a system of government inspection maintained quality, and the efficiency of catching was enhanced in the latter part of the century by the use of larger boats which were decked over, and the substitution of the lighter cotton instead of linen in the nets. While catching was done by drift-net boats big enough to fish in the open sea, it was the practice to land catches daily, so that curing was more conveniently performed than in the Dutch fishery, being the work of thousands of women in the fishing settlements. Further, from being a two-month summer season on the east coast of Scotland, the fishery became an almost year-round one, as the fleets expanded and developed a high degree of mobility, which gave a better utilization of the capital invested in them. They operated from different parts of the coast in harmony with different seasons, including East Anglia in autumn and the West Coast of Scotland in winter. Ultimately the advent of the steam drifter about 1900 rendered the herring fisheries also less at the mercy of wind and tide, and in the peak year of 1907 the Scottish herring cure rose to over 2,500,000 barrels; and at the same time total British landings approached 400,000 tons.

The British herring fisheries were built mainly on providing cured herring for continental markets, especially in Russia and Germany, and the disorganization of the markets in the interwar period, which culminated with the Great Depression caused severe contraction in the British herring fleets. Norway's herring fisheries, too, had been reviving from the start of the century, and in them the more efficient purse-seine was being increasingly used. More important still was the growth of the German catch, which was increasingly by trawl, until their rising landings surpassed the falling Scottish ones in 1937.

Further extensive changes have followed the Second World War, with the continued fall in consumption of cured herring as living standards have risen, the rise of 'industrial' fisheries (of which the catch is used for reduction to meal and oil), and the coming on the scene of new nations as major producers— especially the U.S.S.R. and Iceland. Until about 1960 the main development was the extension of the use of the trawl as a method of capture, when it was largely adopted in Denmark and Sweden and also used by the Russians, although the drift-net had an indispensable place for the faster-swimming herring shoals in summer. In the southern North Sea a big 'industrial' trawl fishery largely for immature herring developed, and much of the Norwegian catch was also used for reduction.

The development of purse-seining for herring in the open sea in the 1960s is an example of the powerful economic impact of a technological innovation—that of the hauling of the net with a power block, coupled to its use from bigger steel boats in the open sea, and the making of larger nets (up to 90 fathoms deep and 300 fathoms long) from lighter synthetic fibres. In Europe, this was pioneered in Iceland, taken up on a big scale in Norway from 1963, and first used extensively in the North Sea fishery in 1965. The result was a sharp increase in what had been the normal twentieth century overall total of 600,000 to 700,000 tons a year to 1,300,000 tons in 1965, although catches have since dropped again. Such spectacular increases in the catch—coming as they did with over-production in anchovetta-derived fish meal in Peru (the world's main suppliers)—dislocated the market outlets and led to a series of restrictions in the Norwegian effort. Even so, purse-seining has largely displaced drift-netting in the open sea and is now the main method of capture.

The modern situation in the herring fishery still includes a considerable consumption of herring as a food-stuff, although now as an item to vary the diet rather than as a staple. It features now more in the form of fresh, chilled, filleted, smoked, or as various preserves, and part of the pattern now is the landing of vessels at foreign ports (see chapter VIII). Herring meal has been used in increasing volume as a feeding stuff in livestock farming, and herring oil has a big range of uses.

The rise of the herring fisheries had important secondary effects, some of them with a large degree of permanence. Right until, and even during, the period of the Industrial Revolution, they played a big role in uplifting regional economies from a near-subsistence level. Thus, under the Hansa, there was considerable seasonal employment around the Sound in the making of 'bückling" (red herrings) and in curing, as well as in fishing and transport of the catch. Although the Dutch cured their catches aboard, the scale of their efforts was such that a leading activity of nearly all the main coastal towns lay in the fitting out for the fishery and the disposal of the catch. In the Swedish fishery the variety of materials assembled at the fishing bases included salt from France and the Mediterranean, hemp from Russia, grain from the Baltic as well as barrel staves from Swedish timber.[24] In south-west Norway, the nineteenth century herring fisheries have been recognized as initiating a new faster tempo of economic growth, which had a multiplier effect on employment and which later provided capital for other enterprises.[25] At the same time in Britain herring fisheries brought a century of economic growth to many coastal settlements, especially on the eastern seaboard of Scotland. If the fisheries have been displaced generally in western Europe in the twentieth century by other growth sectors in the economy, they must now be playing an important role in the ports of the U.S.S.R.

Trends in population and settlement patterns have also been influenced by the herring fisheries. The growth, and in some cases the founding, of many coastal towns can be traced to them. Thus the herring on the coat of arms of Lübeck—the leading town of the Hansa—still shows its old role, while Amsterdam is said to have a foundation of herring bones, and the same might be said of Rotterdam, Enkhuizen and many other Dutch towns. Such towns as Torviken and Marstrand in Sweden, Ålesund and Florö in Norway; Great Yarmouth, Wick, Peterhead, Fraserburgh in Britain also owe their present sizes to the herring fisheries; while Haugesund in Norway and Lerwick in Shetland can truly be quoted as created by them. Also a part-legacy of the fisheries is the close texture of rural settlements patterns in Western Sweden, south-west Norway and the coasts of the crofting counties of Scotland, as there were increases in population in all these areas in the eighteenth and nineteenth centuries, together with subdivision of farm holdings as many households became partly dependent on the cash yield of boat crews or shore workers: these included women as well as men, and there was a great deal of seasonal migration to the bases from which the fisheries were conducted.

While the herring fisheries have provided some rich rewards, they have also been fickle, and the variability in their precise location and their yield from year to year have always led to difficulties in harmonising supply and demand, and have led on many occasions to dangerous financial speculation.

THE COD FISHERIES

In Europe, cod has been by far the most important species of white fish in both historical and modern times. This derives from its abundance, its food value and its suitability for preservation, and it has been exploited from all the seaboard countries of Western Europe without exception; it does not occur, however, in the Mediterranean.

There is evidence from various parts (especially Scandinavia) of cod—with other white fish—being exploited in the Neolithic period, and this generally meant the use of some sort of boat and fishing with hook and line. For many centuries inshore fisheries from open craft must have predominated, and most of them were doubtless in the context of local subsistence. Records of the Viking period, however, suggest that commercial cod fisheries were then beginning to be of significance on the Norwegian coast. By the twelfth century the Basques were venturing out to sea in bigger, decked craft to catch cod,[26] and were splitting and salting them to produce 'bacalao' or the hard-dried 'haberdines'. The cod was well suited to such preservation methods as it grows to a large size, which minimizes the work involved in splitting and salting, and this work could be performed aboard boats, although drying had to be completed on shore. A further advantage was that it needed no barrelling, as did the fatty pelagic fish like herring. Bacalao especially became one of the most important items in the trade in food-stuffs of Southern Europe, and commerce in it from Iberia developed with all parts of the Mediterranean.

Early developments in commercial cod fisheries also occurred in Norway, although the catching here was inshore, and the product was stockfish—cod split and dried without salting—a method of preservation only possible in the low temperatures of the north. By the twelfth century, Bergen was an organizing centre, although most of the fishing was further north, especially in the Lofoten area. Norwegian stockfish were the other main source of preserved cod for Medieval Europe.

While the Basques appear to have been the pioneers in deep-water fisheries, there is Medieval evidence for the involvement of different national groups in them. There were ventures into Icelandic waters as early as the twelfth century, although working of these richer northern grounds in any strength had to wait till later; but by the fifteenth century vessels from bases around the southern North Sea were making regular trips to both Iceland and Lofoten.[27] Of signal importance in the following century was the rapid development of the Grand Banks of Newfoundland, where the obvious wealth of the fisheries attracted the interest of all the sea-powers of Western Europe after Cabot's discovery. As a result, by the seventeenth century the salt cod market was dominated by distant-water supplies.

In the initial development of the fisheries at Newfoundland,[28] much of the activity was with small boats inshore, employing seines and hand-lines: salting and drying for preservation were easier on beaches, and one of the main functions of the bigger boats that crossed the Atlantic was to bring the crews and their stores, and to transport the catch. The same sort of system had also been employed previously on the coasts of Iceland and the Northern Isles of Scotland. Fishing on off-shore banks was also practised from the start, however, and an example of this is the Breton boats, which were up to 200 tons with crews of up to twenty-five. Even such vessels seem to have generally employed hand-lines, and with these a fisherman might catch up to 200 fish in a day. From these boats also come records of an early example of the method of division of the proceeds. The share got by the fishermen depended on the processing method employed: wet salting aboard demanded less work, and from it the crews got one-quarter of the proceeds, while in hand dry-salting the fish had to be taken ashore and spread on beaches to dry: here the men got one third of the total, while one-third went to the vessel-owner and one third to the merchant who fitted out and provisioned the boat. Bait was secured by catching capelin and herring, and using cod entrails. Cod livers were boiled down for oil. By the late sixteenth century, in several of the countries involved, vessels might be equipped with dories for use in the open sea, which allowed the working of a wider area; and in the seventeenth century the use of long-lines became a main method of fishing, which also made for a higher level of efficiency.

Vessels fitted out for the distant-water fisheries usually made two or three trips per year, each lasting up to four months. Salt was inevitably a basic material, and was carried across the ocean by the vessels themselves. Home supplies of 'solar' salt were available to those of France, Spain and Portugal, while those from English ports like Bristol generally called at Portuguese ports for supplies before crossing to Newfoundland. A main objective was to get the first supplies of the year back for Lent markets, and the higher prices then obtaining caused some masters to bring back their boats only half-full. As well as becoming an important food-stuff in European countries, salt cod became a basic item in naval stores and indeed helped to bring about the achievements of the Great Age of Discovery, as well as provisioning the vessels themselves which set out for the Grand Banks.

The orientation of the trade in fish after the development of large-scale distant-water fisheries varied for the different nations involved. In the sixteenth century, activity was dominated by the French and Portuguese, and as well as catering for their home markets they developed an export trade, especially to England and the Mediterranean. The French fleet, however, was always principally engaged in supplying the Paris market, and when English sea-power grew in the seventeenth century England became a main supplier to Spain and the Mediterranean, and Bristol became a European entrepôt in the salt cod trade, gathering supplies from Newfoundland, Iceland and Norway for distribution. There was also a trade of considerable volume with the developing plantations of the West Indies and the mainland of North America. The growth of European settlements in North America, and the attainment of independence by the United States, meant that from the latter eighteenth century the main interests in the fisheries of the western Atlantic passed from the west coast of Europe to Canada and the United States; but the European nations have retained an involvement in them till the present time.

If the most important developments of the Mercantilist period in cod fisheries were in the north-west Atlantic, there were important, if lesser, developments in the fisheries of the north-east part of the ocean in the eighteenth century. In Norway, in particular, there was a growing volume of production of klippfish (salt cod) to supplement her traditional product of air-dried and unsalted stockfish.[29] This new development was based especially on enterprise and organization in the towns of Ålesund and Kristiansund, which were collecting points for catches from most of the west coast. Norway made inroads into the Mediterranean markets at the expense of other European powers, and in the eighteenth century there was also some development of the salt cod (and ling) trade from Scotland. In the nineteenth century there was renewed effort on Iceland and Faroe grounds from British and Norwegian bases, while Iceland herself had a rising export. In the days of sailing ships, a feature of operation from English ports was the use of the Orkneys and Shetlands as landing and turn-round points in the fishing season.

For the great part of the history of the cod fisheries, it was essential to dry and salt the product for preservation, and it was mainly eaten by the poorer classes. From at least the seventeenth century, however, there has been systematic catering for a wealthier sector of the market, when the leading sea-faring nation of the Dutch conducted part of their fishery in well-boats which brought the fish back alive from the North Sea grounds.[30] The technique was also used in Britain in the eighteenth century, when cod was brought back with a variety of other white fish; and in the latter eighteenth and the nineteenth centuries well-smacks were fitted out in ports in eastern England to fish as far afield as Orkney and Shetland.

The nineteenth century growth of trawling was to introduce a new intensity of operation in the cod—and other—fisheries, and as a result cod became in the early twentieth century the main food fish of Europe, when it displaced herring; at the same time Britain became again for a period the leading nation in the fishery, and in the twentieth century a high intensity of exploitation by trawl has spread out from the North Sea to cover all the stocks of the North Atlantic, including the Barents Sea and Greenland waters. Further, from being almost entirely salted before the First World War, most of the catch is now delivered to market in iced or frozen form, and fresh or frozen cod is now the main fish item in the diets of Europe.

THE MODERN SITUATION OF EUROPEAN FISHERIES

The pattern of fisheries exploitation from European bases is thus the end product of centuries of growth, and in the twentieth century activity has been geared up to an unprecedented pitch. At present there is still a large element of traditional ways to be seen, with the operation of crofter-fishers at various locations around the coasts from Norway to Greece, and with methods of preservation which still include drying, salting and smoking and traditional marketing channels at long and short range. Everywhere, however, the accent is on development, on improvement and rationalization in catching, processing and marketing, although irregularities in supply provide formidable obstacles to detailed planning. While activity on some scale is still dispersed around the coasts of the continent, and indeed on many inland waters, forces of concentration have been progressively increasing over the modern period, and activity is much dominated by major ports like Hull and Grimsby, Bremerhaven and Cuxhaven, Boulogne, Vigo, Leghorn, Ijmuiden, Esbjerg, Göteborg, while Murmansk is now said to be the world's largest fishing port. Another notable trend has been the decline in importance of seasonal fisheries, which formerly accounted for the main efforts: big capital investment entails that the fleets and marketing system remain in operation all the year. This century especially has seen much increase of effort in the scientific study of sea fisheries, and their biological basis is now much better understood, although much is still far from clear. It is only within the last twenty years, however, that fishery economics has received any detailed study.

REFERENCES

1 J. G. D. Clark, *Prehistoric Europe: The Economic Basis*, 1952, pp. 84–90.
2 F. Bartz, *Die Grossen Fischereiräume der Welt*. Band I, *Atlantisches Europa und Mittelmeer*, 1964, pp. 359–62.
3 C. L. Cutting, *Fish Saving. A History of Fish Processing from Ancient to Modern Times*, 1955, pp. 18–24.
4 C. L. Cutting, *Ibid.*, p. 25.
5 F. Bartz, *Ibid.*, p. 71.
6 C. L. Cutting, *Ibid.*, p. 40.
7 C. O. Sauer, *Northern Mists*, 1968, pp. 64–68.
8 A. Beaujon, *The History of the Dutch Sea Fisheries*, 1883, p. 431.
9 P. G. Borisov, *Fisheries Research in Russia: A Historical Survey*. English Edition, 1964, p. 6.
10 D. H. Cushing, *The Arctic Cod*, 1966, pp. 1–24.
11 A. Hardy, *The Open Sea* (vol. II), 1956, pp. 146–47.
12 G. L. Alward, *The Sea Fisheries of Great Britain and Ireland*, 1932, pp. 193–265.
13 P. G. Borisov, *Ibid.*, p. 72.
14 F. T. Christy (Jr.) and A. Scott, *The Common Wealth in Ocean Fisheries*, 1965, p. 40.
15 W. R. Mead, *An Economic Geography of the Scandinavian States and Finland*, 1958, p. 175.
16 A. E. Christensen, 'Scandinavia and the Advance of the Hanseatics', *Scand. Ec. Hist. Rev. V.*, 1957, p. 110.
17 C. L. Cutting, *Ibid.*, pp. 57–60.
18 C. L. Cutting, *Ibid.*, p. 66.
19 A. Beaujon, *Ibid.*, p. 360.
20 C. L. Cutting, *Ibid.*, p. 368.
21 G. Utterstrom, 'Migratory Labour and the Herring Fisheries of Western Sweden in the Eighteenth Century', *Scand. Ec. Hist. Rev.*, VII, 1959, pp. 19–27.
22 R. Östensjö, 'The Spring Herring Fishery and the Industrial Revolution in Western Norway', *Scand. Ec. Hist. Rev.*, XI, 1963, p. 145.
23 M. Gray, 'Organisation and Growth of the East Coast Herring Fishing 1800–1885', in *Studies in Scottish Business History*, ed. P. L. Payne, 1967, pp. 187–216.
24 G. Utterstrom, *Ibid.*, p. 5.
25 R. Östensjö, *Ibid.*, pp. 150–54.
26 C. O. Sauer, *Ibid.*, p. 64.
27 E. Power and M. M. Postan, *Studies in English Trade in the Fifteenth Century*, 1933, pp. 173–74.
28 H. A. Innis, *The Cod Fisheries*, 1940, pp. 11–26; 43–60.
29 C. L. Cutting, *Ibid.*, p. 145.
30 A. Beaujon, *Ibid.*, p. 438.

CHAPTER V

The Yields of the Different Sea Areas, and the Countries Involved in Exploitation

The pattern of exploitation of fishing grounds by the fleets of Europe is primarily governed by the productivity of the grounds and their distance from fisheries bases. The outward extension of the intensively worked area which began last century has continued to the point that it now covers virtually the whole North Atlantic, together with West African grounds, the Mediterranean, Black and Caspian Seas, and inland European waters. Yields from the longer exploited nearer grounds have tended to stabilize, and the greatest rates of expansion since the Second World War have occurred on the distant-water grounds of the North-West, East-Central and South-East Atlantic. This increased expenditure of effort on distant-water grounds has involved most of the major fishing nations like the U.S.S.R., Spain and Britain, but Denmark and the Netherlands have achieved expansion by more intensive exploitation of nearer waters.

An examination of the trends in the major sea-areas recognized by international conventions and by F.A.O. will show the trends in greater detail.

I.C.E.S. Area (International Council for the Exploration of the Sea: North-East Atlantic). In this area (Figure 5), expansion since the Second World War has not been sustained, largely because there has been limited scope for further intensification. The yield from this area was already approaching 3 million metric tons in 1913 and by 1938 was well in excess of 4 million. Increase continued after the Second World War, but the rate is not clear as the rapidly expanding U.S.S.R. catches do not appear in the statistics until 1955, when the total was nearly 8 million metric tons, and it has fluctuated around this value since. Big herring catches with more modern catching methods increased the total to over 10 million metric tons in 1966 and 1967, but it is now declining. Within this most important area there are now signs that the maximum sustainable yield has been passed for some important stocks. The total demersal catch has decreased some 20 per cent. since 1956, and that of the supremely important cod halved in the decade 1956–65.

All the major fishing nations depend heavily on the I.C.E.S. area, and the only coastal countries not directly involved in its exploitation are those of south-eastern Europe.

I.C.N.A.F. Area (International Convention for North-West Atlantic Fisheries). This has been a source of fish for the Western European nations since the sixteenth century, although in the earlier part of this century the more conveniently located nations of the U.S.A. and Canada, together with the former dominion of Newfoundland, dominated the fisheries. The catch has greatly expanded from 1·8 million metric tons in 1954 to over 3 million by 1965 and over 4 million in more recent years. France and Portugal have maintained their large share in the catch, and those of several of the European nations—especially the U.S.S.R.—have been rapidly increasing so that they now take over two-thirds of the total, and the U.S.S.R. challenges Canada for leadership in the fisheries.

The Mediterranean. Estimates of the yield from the Mediterranean together with the Black Sea, in both of which the European nations are the leading exploiters, show a post-war increase of about 50 per cent., although since 1960 the catch has stabilized at around 1 million metric tons.

European Inland Waters. Fish from fresh inland waters have approximately doubled in Europe (excluding the U.S.S.R.) to 200,000 metric tons since the war; despite mounting river pollution, the growth of fish farming has promoted expansion. Around half of this total may come from fresh waters in the U.S.S.R., although here pollution, together with interference with rivers for power and irrigation is reducing the catch. The same causes have reduced the catch from the Caspian Sea by about one-half of its 1913 total of 500,000 metric tons.[1]

East Central & South-East Atlantic. The yields from these grounds have increased proportionately more than those of the North Atlantic in post-war years, and the European nations have participated especially in those from the former because of its more convenient location. The U.S.S.R., Spain, Portugal, France and (to a lesser extent) Italy have all increased their catches from the formerly little exploited East Central area, and they have been in competition with the African coastal nations and with Japan. This area yielded only c. 100,000 metric tons in 1953 and c. 300,000 in 1962, but a subsequent vigorous expansion raised the 1967 total to 1,600,000 metric tons. In the South-East Atlantic, the trebling of production between 1953 and 1967 to 2,500,000 metric tons is due primarily to the development of the South African industrial fishery, but the U.S.S.R., Spain and West Germany also exploit this area.

CATCHES AND TRENDS WITHIN THE I.C.E.S. AREA

Statistics are available in greater detail and over a longer period for the I.C.E.S. area of the north-east Atlantic, and these allow its importance for the European nations to be examined in greater detail.

Within this area, the most important single part is that of the North Sea, because of its short distance from the bases of most of the major maritime nations as well as of its productivity: the annual yield is now c. 2 million metric tons. The Arctic grounds of Iceland and the Norwegian Sea, which involve distant-water operations for the majority of nations, each produce well over 1 million metric tons per year, and the Barents Sea has attained this figure in some years. Several of the near-water areas now yield around 400,000 metric tons— the Baltic, the Bay of Biscay and Portuguese waters; and the yield of the distant-water grounds of West Greenland are also at this level. Other areas contribute lesser amounts to the overall I.C.E.S. total, although on several of the smaller near-water areas (see below) the yield in relation to their sizes is very high.

Varying trends are apparent in the catches of different areas. On several of the longer-exploited grounds, there has been no consistent increase, but rather fluctuations around a mean. This applies to the North and Norwegian Seas, Icelandic grounds, the west and south coasts of Ireland, Faroe Grounds, the Baltic, the Sound and the Belt. There have been significant increases on some of the nearer grounds, although production from most of these now appears to be approaching an equilibrium. The doubling of the catch from Portuguese waters between 1951 and 1967 has been largely due to more efficient catching methods—particularly trawling—supplementing or superseding simpler small-scale techniques. The yield from both the Irish Sea and north-west Scotland has increased by a similar proportion over the same period due to more intensive fishing. In the Kattegat and the Skagerrak the doubling of the catch has been almost entirely due to the expansion of the trawl fishery in Denmark and Sweden for herring, sand-eels and Norway pout for reduction.

On the northern margins of the exploited areas, catches have fluctuated markedly. The sea areas involved include Spitzbergen and Bear Island, East and West Greenland and the Barents Sea, and a combination of circumstances contributes to the irregular yield. Hydrographic factors affect the location of stocks from year to year, and conditions of weather and drift ice the catching effort expended; in addition there is the modern over-exploitation of the Arctic cod, which provides the great bulk of the catch in these areas. The most prominent result of this has been the marked fall in catches in the Barents Sea since 1956.

The broad relationships between the location of fisheries bases and the areas they exploit can be seen within the framework of national fishing activities. The areas worked by different countries are related both to the distance to fishing grounds and to the demands of their own particular markets. All countries exploit their own inshore waters, although at different levels of intensity; and the proportion of effort expended on more distant grounds varies much more. Thus the catch of the Netherlands comes almost completely from the North Sea, that of Finland from the Baltic, and the yield of the Bay of Biscay is almost completely shared between France and Spain. The Danish and Swedish catches come largely from local waters including both Baltic and North Sea and the area between. The big Icelandic catch also comes almost entirely from home waters, and in south-east Europe, Yugoslavia Bulgarian and Roumanian fleets are all confined to local grounds.

Other sea-board nations exploit distant waters to some degree. Of the other major fishing nations, the proportion of the catch coming from distant grounds is least in Norway, because of the large catch from the rich local grounds of the Norwegian Sea, although it exploits all the northern grounds in some degree. The distant-water fleets of Britain and West Germany also range over all the northern waters, and more than one-half the West German catch now comes from these; but the distant-water effort of Belgium is limited to Iceland waters. Faroese catches are dominated by landings from the north-west Atlantic (especially West Greenland) and these now account for some two-thirds of the total.

While most of the Spanish and Portuguese catches come from local waters, a substantial proportion in each case comes also from distant-water grounds in the North-West and East Central Atlantic. For Portugal, about one-third of the total catch is cod from the North-West Atlantic, and a smaller proportion comes from grounds off north-west Africa and the Azores. Spain takes nearly one-quarter of her catch in cod from the North-West Atlantic, and about one-sixth from grounds off north-west Africa, including those around the Canary Islands. France also expends most effort on distant-water working in the North-West Atlantic, but French fleets range exceptionally widely, from Spitzbergen in the north to West Africa in the south. Italy and Greece are involved to a lesser extent in long-range operation, with trawlers working the north-west African Shelf off Mauretania.

The leading nations in distant-water operation are now those of Eastern Europe; this is their only means of becoming major participants in sea fisheries. Upwards of 90 per cent. of the Polish catch now comes from outside the Baltic, and most of that from distant-water grounds in the Norwegian Sea. East Germany is also developing in the same direction, and of the total Russian catch of over 5 million tons, around 2 millions now come from distant-water Atlantic grounds.

An example of extension of distant-water operation in the

period since the Second World War is given by the Portuguese demersal catch from grounds off north-west Africa, on the Mauritanian and Senegal coasts (Figure 4). This shows the division of fresh demersal landings between local and distant waters, but does not include the salt-cod from the North-West

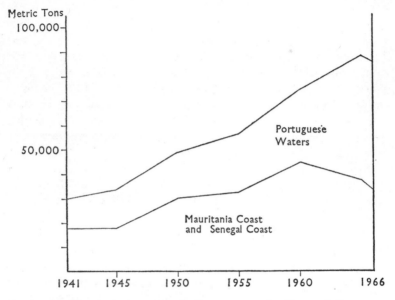

Figure 4. Trends in near- and distant-water landings of Portuguese trawlers, 1941–1966.

Atlantic which actually comprises most of the national demersal catch. The catch from the north-west African grounds increased by two and one half times in the years between 1945 and 1960, and during this period consistently accounted for around 60 per cent. of the landings of demersal fresh fish. In those years the catching rate of the trawlers which worked the distant grounds was from three to four times that of those fishing home waters, but the resulting thinning of the stocks was followed by a decline to around double the rate in home waters. Since 1960 the distant-water catch has declined, while more intensive exploitation of Portuguese waters has led to a continued increase in their yield: they now account for around 60 per cent. of the total.

CATCH LEVELS PER SQUARE KILOMETRE IN THE OCEANS

The catches in different areas and by different countries are related basically to the levels of biological productivity discussed in Chapter III, and sufficient data are available to

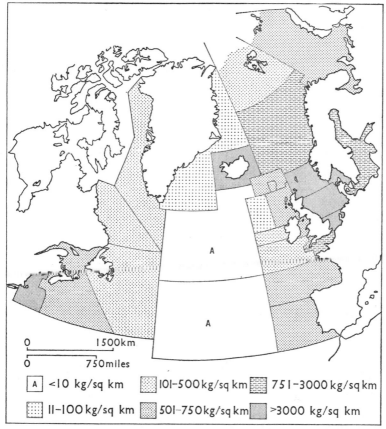

Figure 5. Average yields in sea areas of North Atlantic for I.C.E.S. (North East Atlantic) and I.C.N.A.F. (North-West Atlantic) subdivisions.

show the broad variations in yield over the I.C.E.S. and I.C.N.A.F. areas. This can be expressed in average kg./sq.km., and this varies from zero in deep water of the mid-Atlantic to over 5,000 kg./sq.km. on the most productive grounds on the shelf; and outside these areas in the exceptional conditions of the Sea of Azov, the yield has been computed at 7,890 kg./sq.km. The level on many exploited grounds, however, is only around 500 kg./sq.km., and this is about the average for the Mediterranean. A complication in using such figures of productivity is that many of the sea areas extend outward from the shelves into deep water, while the catches are largely concentrated on the shelves. Even so, the pattern shown in Figure 5 indicates in a generalized way the productivity of different areas, although it is modified in some important details.

The sea area with the highest average productivity is that of the Skagerrak and Kattegat, where it is of the order of 4,400 kg./sq.km., while that of the North Sea is 3,270 kg./sq.km., and Icelandic grounds 3,020 kg./sq.km. Whereas the former two areas are almost wholly shallower than the 250 fathoms limit of trawling, more than a third of the Icelandic area is not; and the yield here in relation to workable shelf is in the region of 5,000 kg./sq.km. The Norwegian Sea area and Portuguese Waters have much higher proportions of the less productive deep water in them, and in the former the yield on the shelf area exceeds 4,000 kg./sq.km. and in the latter it is highest of all the open-sea waters and exceeds 6,000 kg./sq.km.

In the North-West Atlantic area the yields of the grounds are generally lower than in the North-East, although this is partly because they are more selectively exploited—especially for cod—and the more plentiful pelagic species constitute a small element in the catches. Only the relatively small sub-area 5 of the I.C.N.A.F. Convention which is off New England, do yields approach the better ones on the other side of the Atlantic: here the average yield of 3,430 kg./sq.km. consists mainly of silver hake and haddock, although it is little worked by European nations. Productivity in other sub-areas varies from 300 to 650 kg./sq.km., and even when catches are related to shelf area the levels do not much exceed 2,000 kg./sq.km.

THE CONSEQUENCES OF DISTANCE IN THE EXPLOITATION OF DISTANT-WATER ARCTIC GROUNDS

The exploitation of distant-water grounds inevitably means that a big proportion of the time at sea must be spent in steaming to and from the grounds; and this has to be compensated by higher rates of catching.

From the main bases around the southern North Sea, Icelandic grounds are at a range of around 1,000 miles, and require seven or eight days' travelling time for the round trip; at 1,500 miles the Barents Sea requires 10 days, and West Greenland and Newfoundland grounds at 2,000 to 2,500 miles require over 14 days. Russian trawler fleets from Murmansk have spent the total of a month making the passage to and from Newfoundland waters and the average distance from base at which the long-range vessels of the U.S.S.R. operated in 1968 was about 3,000 miles. Most distant-water vessels still preserve their catch in ice, which means that they must return to port within about three weeks of beginning fishing; in effect this means they use from 30 per cent. to 60 per cent. of their time at sea in making the passage to and from the grounds. Although vessels engaged in freezing and salting can spend longer on the grounds, 45 per cent. of the time at sea has been required in travelling time by Russian vessels operating on a fleet system with mother-ships on the Grand Banks[2] and in the 1960–65 period, for Russian B.M.R.T. vessels, the proportion of time actually spent on fishing ranged between 39 per cent. and 45 per cent.[3] In addition to the time required on the passage to and from distant grounds, fishing time is particularly likely to be lost on the Arctic grounds through adverse weather conditions. There can also be complications in trans-shipment at sea by catchers to processing and carrier vessels, although the big modern ships in use can come along side each other in conditions as severe as wind force 6 (32 knots).

The greater rewards which are the incentive for the working of Arctic grounds can be illustrated by the average catches of British trawlers for 1950.[4] In that year the grounds of the Norwegian Coast, Spitzbergen and West Greenland all yielded over 100 tons per 100 hours fished, which was more than fourteen times the yield from the North Sea (7·1 tons per 100 hours) and the English Channel (7 tons per 100 hours). In the same year the rate of catch from Faroe grounds was more than four times that of these nearer grounds, and that from Iceland nine times. While allowance must be made for distant-water trawlers being bigger and more powerful than those working local grounds, the much greater yields from Arctic grounds are evident.

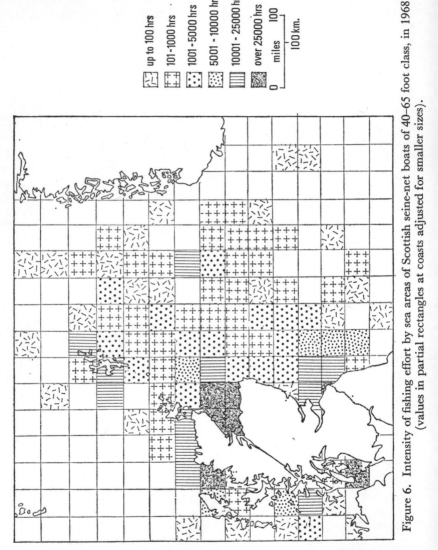

Figure 6. Intensity of fishing effort by sea areas of Scottish seine-net boats of 40–65 foot class, in 1968 (values in partial rectangles at coasts adjusted for smaller sizes).

up to 100 hrs
101-1000 hrs
1001-5000 hrs
5001-10000 hrs
10001-25000 hrs
over 25000 hrs

While distance of fishing grounds from base conditions the operations of all fishing fleets, its effects can be seen most clearly with smaller craft, as they have not the carrying capacity for either stores or catches to give them the freedom in range of movement enjoyed by bigger vessels. The distribution of effort in demersal fishing around Scotland for a sample year (1968) is shown in Figure 6 for seine-net boats of the 40 to 65 feet class. The effort expended on demersal fish shows a clearer pattern than that on pelagic fish, which are much more sporadic in their occurrence. Although this is a section of what is classed as the 'inshore fleet', they are capable of operating at ranges of 200 or 300 miles from base. The statistical rectangles for which numbers of fishing hours are shown are the standard ones of I.C.E.S., and measure ½° in latitude by 1° in longitude.

The general pattern is that of a progressive decrease in effort off-shore, and the great part of the time in fishing is spent within 50 miles of land. There is not an even distribution of effort in the inshore zone, and the effort is most concentrated on the better grounds of the Moray Firth, the Minch and the North Channel, in all of which there were over 25,000 hours fishing. Distance offshore does not preclude a concentration of effort on the Coral and Ling Banks, which lie around 150 miles east of Orkney, and on which there were 17,600 hours fishing. To a lesser extent effort on the Viking, Bergen and Great Fisher Banks which lie nearer Norway than Scotland was noticeable. The plotting of monthly data would show a more pronounced decrease in effort with distance off-shore in winter, as vessels of less than 65 feet concentrate more on nearer grounds in the part of the year when danger from gales is higher.

SEASONAL VARIATIONS
IN OPERATIONS & LANDINGS

The general effects of seasonal variations in fisheries have been discussed in Chapter II. Such annual fluctuations in yield pervade all fishing effort to some degree, although they are generally most marked in pelagic fisheries. While no figures are available to show total effort and yield of European fleets for different sea areas by months, the seasonal fluctuations can be illustrated by monthly statistics from individual countries.

For Danish seiners, the leanest period of the year is the first quarter, and only 10 per cent. to 15 per cent. of their number of hauls is made then; around 35 per cent. of their effort is made in the second and third quarters, and the balance in the last. By contrast Danish trawlers make nearly half their hauls in the

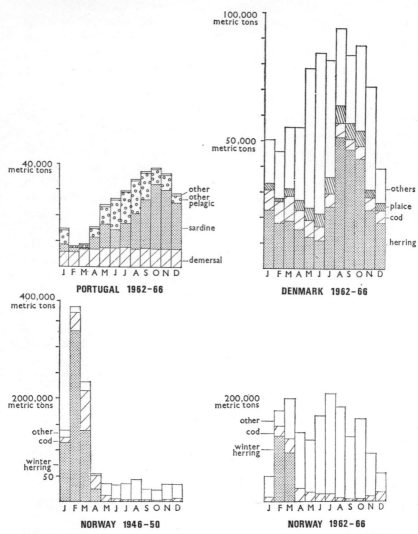

Figure 7. Seasonal fluctuations in landings in Portugal, Denmark and Norway.

spring quarter, and the summer is the minimum period with around 15 per cent. of the effort. Examples from Scotland show that effort of trawlers operating in the North Sea from Aberdeen approximately doubles in the peak summer months (May to August) from the January minimum; but for seiners operating in the Moray Firth the number of hours fishing doubles in January (in the cod season) from the minimum which extends from May to November.

Monthly details of landings are also published by several countries, and these are illustrated by Figure 7, which shows monthly averages for Portugal, Denmark and Norway for the five-year period 1962–66, and also for Norway for the period 1946–50. Such fluctuations are related basically to the seasonal availability of fish, but the Norwegian case also shows the influence of big long-term stock fluctuations. These countries also show the effects of different stages of economic development, as greater efforts are made in Denmark and Norway to minimize seasonal fluctuations by longer-range fleets.

In the Portuguese landings, the 'low' period in the first four months of the year is obvious, and the maximum landings in October are nearly five times those of the February minimum. This difference would be more than twice as great if the 'bacalao' (salt cod) landed from the Grand Banks, were included. This is landed in autumn and comprises about one-half the demersal catch, or one-fifth of the total for all species. The official monthly statistics, however, include only fish landed fresh. It is notable that demersal landings are substantially constant throughout the year. This reflects their unbroken availability, but also the fact that more than half of them are taken by a mobile trawler fleet which operates off north-west Africa as well as in Portuguese waters. Seasonal variations are very much due to the periodicity of the pelagic fisheries, and outstanding is the concentration of the sardine fishery in the latter part of the year. While these fluctuations present problems in gearing the capital equipment employed in the fisheries to the needs of different seasons, the overall variations are scarcely as great in economic terms as the landings tonnages suggest. The unit values of demersal fish, and of molluscs are considerably greater than those of pelagic species, with the result that total landings values at the October peak are only about twice those of the February minimum. Portugal has already decreased the seasonal fluctuations in her fisheries by developing middle- and distant-water trawling. they are likely to decrease further in future.

The Danish monthly landings show the most usual situation of the seaboard countries of Europe outside the Mediterranean in having a relatively high level of catch in all months, but with a maximum in the warmer part of the year. This is due mostly to the fact that this is the main herring season of the North Sea, but also to the fact that demersal landings usually increase at this time of year largely because of longer daylight and better weather for operations. The peak tonnage of August is almost two-and-one-half times that of the December minimum. The total monthly tonnages landed in Denmark give no clear indication of values, however, as a large proportion at all seasons goes to supply the big reduction industry established in connection with Danish farming and with export. This can account for 80 per cent. of the landed weight in a given year, but almost never accounts for more than one-third of the value; it consists mainly of herring, Norway pout and sand-eels, and the last named are useful for maintaining landings in the lean spring period. In actual fact, while there is a tendency for total landed values to be greater in the months with higher landings, this is not markedly so, and in any case is as much due to the greater landings of the very valuable plaice at this time as it is to the big volume of fish for reduction. Total values in the colder part of the year are also maintained by greater catches of cod and the high-value salmon at this time. Plaice, cod and salmon together comprise more than one-third of all Danish landings by value.

The Danish industry does represent a relatively successful adaptation to modern demands. Although many of the craft are small, mechanisation has advanced and the landings per vessel have increased dramatically since the Second World War. This is one of the few unsubsidised fishing industries in Europe.

In Norway the post-war changes in the pattern of seasonal landings are striking. The movement away from the old pronounced winter peak in landings to a more constant seasonal distribution is related partly to national policies of economic rationalisation, by promoting full-time seasonal operation with larger craft. The change is also due, however, to serious failures in the main traditional fisheries of the Arctic cod and (more especially) the winter herring. In the period 1946–50 in the traditional fisheries landings were high and conditions substantially prosperous: at this time more than 70 per cent. of all Norwegian landings were made during the first three months in the year. While even now these fisheries make an important contribution to total landings, there has been expansion of off-shore operations, from which much of the yield comes at other times of year: these include herring fisheries in the North Sea and at Iceland, and distant-water trawling and lining for white fish. There are now lesser fluctuations in the pattern of annual landings, and there is a second peak in mid-summer which is as important as that in the early months of the year. Figure 7 shows the former overwhelming importance of the winter herring fishery, and the pronounced late winter peak in cod landings; now the only low point in cod landings is in autumn, while the winter herring yield has dwindled to a fraction of its former level, and landings of a variety of other species have much expanded. Formerly the winter herring and Arctic cod between them accounted for about one half the value of landings, but now for only about one-fifth.

While the same tendency as in Norway towards smoothing the curve of annual landings exists in other European countries, fluctuations persist in all in some degree. In the 1920s German landings in the peak months of September and October were up to four times those of the minima in June and July; the difference is now less than double. German distant-water trawlers in recent years, by exploiting a variety of grounds at different seasons, have reduced the amplitude of variation in their yield to around 40 per cent. for different months. There is a fairly constant supply through the year from Iceland grounds, but with something of a low period in summer, and this summer 'low' is more pronounced at Faroe. Operation at the Norwegian coast is largely confined to the main cod season in spring, and effort in summer is diverted to West Greenland, where the fishery continues into autumn. There is also a regular fishery on George's Bank off New England from July to December, and the trawler fleet also works grounds in the North Sea and off the west coasts of Britain in late summer and autumn. The wide-ranging fisheries of the

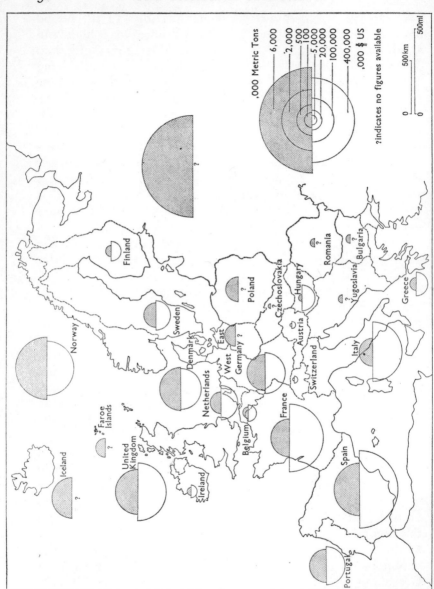

Figure 8. Tonnage and value of landings for 1967 in the countries of Europe.

U.S.S.R. have been among the most successful of the European nations in minimizing seasonal fluctuations in landings. In 1929, 47 per cent. of landings were in the second quarter of the year, with only 5 per cent. in the first quarter, and the remainder being fairly evenly divided between the third and fourth quarters; but by 1965 the proportion for each of the four quarters was within the range of 20 to 30 per cent.[5]

In Italy, the landed weight varies from a summer (May-July) maximum to a winter (January and February) minimum, and the monthly tonnages in summer run at around twice those of winter. These fluctuations are almost entirely due to variations in the availability of pelagic species—particularly anchovy and sardine, but also mackerel. The sardine catches may vary by a factor of seven between minimum and maximum.

VOLUMES & VALUES OF LANDINGS IN THE EUROPEAN COUNTRIES

The pattern of fish landings in Europe is to be explained primarily by the locations of different countries, which have strongly conditioned their opportunities to engage in fisheries. At the one extreme are the leading nations of the Atlantic seaboard (Figure 8 and Appendix), in nine of which landings regularly exceed 500,000 metric tons per year, and at the other the landlocked countries in which it is rare for production to reach 20,000 metric tons. The leading nations include those countries first to develop industrially—Britain, Germany and France; and also included are Norway and Portugal, in both of which farming opportunities are limited by terrain, and in which fishing has always played a major role in the economy. More recent developments in the U.S.S.R., Spain, Denmark and Iceland have brought all of these to the front rank. The U.S.S.R. is now the outstanding fishing nation in Europe, and its catch of 6,080,000 metric tons is now (1968) fourth in the world behind Peru, Japan and China; perhaps 65 per cent. of the landings of the U.S.S.R. are made in Europe. In the U.S.S.R. the plan is to achieve a catch of 10 million tons by 1975; it is envisaged that thereafter the level will stabilise in view of the growing pressure on fish stocks on a global scale. The Norwegian catch now fluctuates around 3,000,000 metric tons, and both Spain and Denmark caught about half this figure in 1968. The catch of Britain, which fluctuates around 1,000,000 metric tons, is the next in size, although Iceland in good years has exceeded that amount; and the catches for other nations for the latest available year are shown in the Appendix.

The value of landings in different countries is by no means directly proportional to tonnage. A number of inter-acting factors determine prices (Chapter VIII); briefly, values of landings depend on the species, the proportion of the catch used for reduction, the competition at first sales, and also on whether they have to bear the later costs of export. Discussion of values here must exclude the countries of Eastern Europe, for which very few figures are published.

In value of landings, Spain is now the leading nation in Western Europe, and the gross value of her catch exceeded that of France (the next nation) by about 25 per cent. in 1967; but the unit value of the French catch was U.S. $324 per metric ton against that of U.S. $228 for Spain, and this is largely explained by the bigger proportions of the more valuable crustaceans, molluscs, and fresh-water fish in the French catch. Although the Norwegian landed tonnage is double that of Spain, its unit value is much lower, both because of the high proportion used for reduction and of the big export sector, and prices of fish even for human consumption are only about one-half those of Spain. The third nation by value in 1967 was Italy, although it was only ninth by volume, and its high average unit value of U.S. $515 per metric ton is explained by the same factors as the high French value.

The average catch values in most of the fishing nations of the Atlantic seaboard now lie within the range from U.S. $100 to U.S. $350 per metric ton, and within this range there is a concentration of values in the range from U.S. $150 to U.S. $200. Values in Norway, Iceland and Denmark, however, all lie well below U.S. $100 per metric ton; this is very largely due to large proportions going at low prices for reduction, but, in Iceland and Norway especially, it also reflects lower prices at first sales for fish which are later exported. Within the Mediterranean zone, average values are raised because of the high proportions of more valuable species like sardine, anchovy, breams and mullets; in addition to the Italian figure, Greece in 1967 had an average of U.S. $415 per metric ton and Malta of U.S. $500. The highest values, however, are realised in inland countries where the whole catch is from fresh waters: the average for these is in the region of U.S. $1,000 per metric ton, and in Hungary was U.S. $1,425 in 1967. The value of the Hungarian catch is now on a par with that of Sweden, although its volume is less than one-tenth.

There is a great variation in the composition of catches in the different nations of Europe, but broad patterns are discernible. More than twenty groups of species are detailed in the international statistics published by F.A.O., but in most countries the majority of these are of little or no consequence. For the most part catches are dominated by the cod group (which includes most important demersal species) and the herring group (in which are most of the main pelagic species). For some countries, however, the groups of flatfish, mackerels, redfish, salmons, crustaceans, molluscs, shads and fresh water species are important.

For all countries of the Atlantic seaboard the cod group features prominently in the landings; for most countries the main species in the group is the cod itself, although cod catches are exceeded by hake in Spain and by haddock in Scotland. With the exception of Sweden, the group comprises at least 25 per cent. by weight of the landings of the countries of the Atlantic seaboard; in several—Belgium, Denmark, West Germany, Norway and Portugal—it is around 50 per cent.; and over 60 per cent. of the Faroese and about 80 per cent. of the English catch come into the group. The general tendency is for the proportional values of the group to be rather less than the above figures, other than in Norway and Iceland, where a substantial part of the catch consists of other species for reduction.

Next in importance is the herring group, and this contains an important regional division within it into the herring and sprat fished of north-west Europe, and the sardine and anchovy of the Mediterranean zone. This group forms a more variable proportion of catches, and Iceland especially has suffered in recent years from the uncertainties of the herring fishery; from being 60 per cent. of a total catch of 1,240,000 metric tons in 1966 herring landings slumped to 25 per cent. of the 600,000 metric tons caught in 1968. Similar, but less violent variations in herring catches have largely been

H

responsible for fluctuations in the Norwegian catch, in which herring have varied from 25 per cent. to 50 per cent. since 1960. The Swedish catch is dominated by herring, and it has formed a more consistent proportion at around two-thirds of the catch; and in Denmark, Poland and East Germany it accounts for about one-third. All other coastal countries from France northwards are involved in herring fishing, but in most now the landings are less than 25 per cent. of the catch, with the exception of Finland, the fisheries of which are dominated by Baltic herring. Herring and sprat are among the least valuable fish, even in Sweden accounting for only about 50 per cent. of the catch value, and in nearly all other countries for under 20 per cent. In the Mediterranean zone, sardine and anchovy usually comprise about 20 per cent. of the catch by both weight and value; and in Yugoslavia they account for 50 per cent. of the catch by weight, but no values are published.

The flatfish group is everywhere of high unit value, although in few countries are they a substantial fraction of the catch by weight. The main landing points are around the shallow waters of the southern and eastern North Sea, where natural conditions make them most plentiful. In Holland they are 20 per cent. of the catch weight but 50 per cent. of the value, and in Belgium 20 per cent. and 40 per cent. respectively. Danish landings at around 60,000 metric tons per year are as large as the Dutch, but the value is only about one-half because of a much lower proportion of the most valuable sole.

The redfish group are of significance in only a limited number of cases. Systematic expansion of the rich redfish fisheries in the north-west Atlantic has led to their proportion in the West German catch rising to around 15 per cent. by weight and 20 per cent. by value; and the U.S.S.R. now has a bigger total catch, although it is only 8 per cent. of the national total. The Spanish catch in this group accounts for about 7 per cent. of the national total by both weight and value, and consists of a great variety of species, including angler fish and red bream. Breams and mullets from this group dominate the Greek catch, and it accounts for about one-half the total landed weight (no value figure is available). The Mediterranean varieties in this group are also important in Italy; they constitute 10 per cent. of the catch by weight and 15 per cent. by value.

While mackerel species are exploited by all European countries, it is only Norway, Sweden, Denmark and the Mediterranean area that they are of notable significance. The Norwegian catch is by far the highest and can exceed 800,000 metric tons, contributing around 25 per cent. of the total catch by weight and 12 per cent. by value.

Although the salmon group has generally a high unit value, it is only rarely that they bulk large enough to be important parts of the catches. Pond trout in Denmark are largely responsible for the value of the group now being over 20 per cent. of the total of national landings, and salmon are a similar proportion by value of the catch in Northern Ireland. By far the biggest volume of landings in this group is the Norwegian catch of the Arctic capelin, which now is over 500,000 metric tons per year. Although this approaches 20 per cent. of the national catch by weight, it is fished for reduction and its unit value is very much less than the species in this group used for food; its value is less than 5 per cent. of the national total. No values are available for the catch of capelin in Iceland—the only other European nation which exploits it on any scale—but in recent years its volume has been around 10 per cent. of the total.

Although widely fished in both the Mediterranean and the Atlantic, tuna are not a major component in the catches of any country. France is active in this fishery, and catches of around 60,000 metric tons are 7 per cent. by weight and 10 per cent. by value. The Spanish catch is of the same order of size as the French, and here it is 4 per cent. by weight and 8 per cent. by value of national totals.

Crustaceans are another group of high unit value, although the restricted quantities of them usually restrict their total value to 5 per cent. or less of national totals, apart from the French catch which is about 10 per cent. by value although less than 3 per cent. by weight. Molluscs include a wide value range: the cuttlefish, octopus and squid which dominate Mediterranean catches are of relatively high unit value, and in Italy the value of the group is 15 per cent. of the national total. In north-west Europe, mollusc catches are dominated by mussels, for which

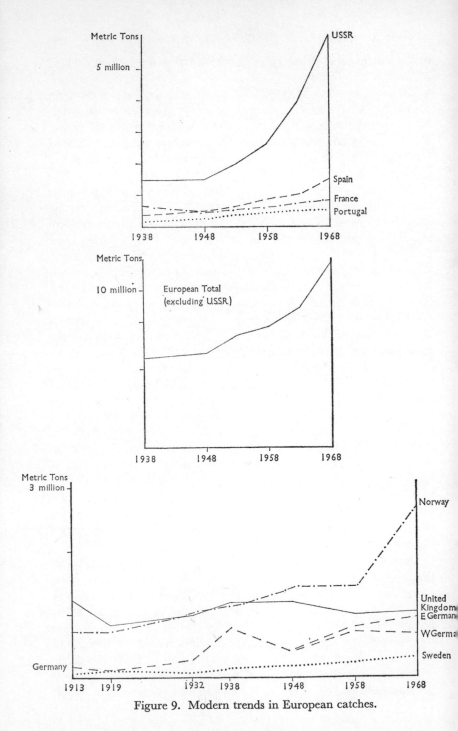

Figure 9. Modern trends in European catches.

values are low: the shallow coastal and estuarine waters of Holland are the most productive area, and they are one-third of Dutch landings by weight but less than 2 per cent. by value.

The shad group is of very limited importance in Europe, apart from in the U.S.S.R., where they total c. 450,000 metric tons per year and now account for 7 per cent. of the catch; this is almost entirely due to the catches of Caspian, Azov and Black Sea sprat, which formerly were a much larger proportion of the national total before large-scale exploitation of the oceans began.

The fresh water catch of the U.S.S.R., although declining, was greater than that of any other European country, and was 397,000 metric tons in 1968; a substantial proportion of this comes from waters in the European part of the U.S.S.R. The fresh-water catch is much more important proportionately to other countries, and not only to the land-locked ones of the interior. In Roumania 70 per cent. of the catch comes from fresh waters, in Yugoslavia over 30 per cent. and in Finland over 20 per cent.: these countries have restricted opportunities for sea fishing, but relatively rich inland waters.

TRENDS IN LANDINGS

The modern period has witnessed a consistent and vigorous upward trend in the European catch (Figure 9) and by 1968 the catch for the continent (excluding the U.S.S.R.) had increased to two-and-one-half times its 1938 level, and five times that of 1913. There was some tendency for the increase to accelerate, with the most rapid growth in the early 1960s, but since 1965 the graph has tended to level out, suggesting that an equilibrium may be approaching. The incompleteness of available statistics prevents a statement of any precise trend in values; in any case the irregular decrease in the value of money, and changes in the relative value of one currency against another, renders this a complex problem, but the nominal value appears to have increased between six and ten times since 1938. The real value, however, has probably increased at a rate approximately parallel to that of the catch weight. Since 1938, the rate of increase in Europe (excluding the U.S.S.R.) has been rather less than the world rate, which trebled up to 1968 when it stood at 64 million tons and this compares with a level which has been estimated at 1·5 to 2 million tons in the mid-19th century.[6] The European increase has been superior to that of North America and of Oceania (which includes Australia and New Zealand); it has been parallel to that of Asia, but since 1938 the U.S.S.R. catch multiplied four times, that of Africa seven times and that of South America more than fifty times.

The rate of growth has shown marked differences throughout the continent. The outstanding increases have occurred in Eastern Europe, and planned growth in East Germany and Poland has led to their catches being multiplied ten times, and the proportion of the Russian catch taken in the open sea has increased by the same proportion. Prominent increases have also occurred in the inland countries and in most of those on the Mediterranean and Black Seas; five-fold increases have occurred in several cases.

Growth has generally been slower in the countries of the Atlantic sea-board, although this is largely because of the much greater scale their operations reached before the Second World War, and consolidation rather than expansion has often been the key-note in development. There is, however, the spectacular increase in the Danish catch, largely associated with the fishery for reduction, and in 1968 landings were more than fourteen times the 1938 figure. Although the Icelandic catch has fluctuated, in good years it has been four times the 1938 figure, and the Norwegian landings have increased around two-and-one-half times and those of Sweden about three times. Spain's prominent position among Europe's fishing nations has been achieved as a result of a three-fold expansion, and Portugal is another of the leading nations that has doubled her catch.

In other countries fronting the Atlantic, which are largely those nations which established industrial economies and capital-intensive fisheries earlier, there has been a greater measure of stability. In recent years catches in the Netherlands and Belgium have been around 50 per cent. above the 1938 level and those of France about 25 per cent. In West Germany and Britain, however, catches are about the same as the 1938 level, and the British catch has actually decreased slightly

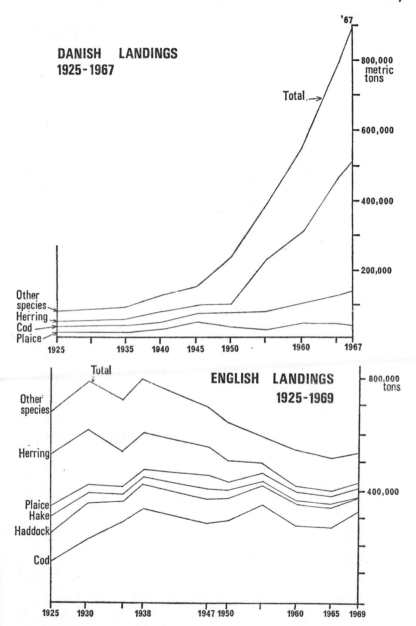

Figure 10. Trends in landings and catch composition in Denmark and England.

since 1913. This would appear to indicate that Britain and Germany earlier attained an equilibrium that several other countries are now approaching: since about 1960 the catches in Portugal, Holland, Belgium, Sweden and Iceland have all tended to fluctuate without any long-term increase. This is the case also in the inland countries of Austria and Switzerland.

While there is a large element of stability in the composition of catches in different parts of Europe, there has been a tendency to expand the effective resource base by the exploitation of new stocks, and in some cases new sea areas. The basic importance of the cod, herring, and sardine fisheries remains, but their relative importance has been reduced.

The systematic expansion of the West German trawler fishery for redfish in the north-west Atlantic is an important instance of the increase in the catch of fish for food, as in the Russian catch of South African pilchard in the South Atlantic. For reduction fisheries, there has been (in addition to herring), the expansion of catches of capelin and mackerel in Norway, capelin in Iceland, and sand-eels and Norway pout in Denmark. For the higher-value sectors of the market, the increase in catch of crustaceans, salmon, trout and some molluscs, has been prominent in many countries, although these varieties are never more than a small proportion of the catch weight.

Trends in the composition of catches, along with the factors that influence them, can best be studied for individual countries, and Figure 10 shows the changing composition of catches in England and Denmark, two countries for which a long series of little broken statistics are available.

The most prominent trends in the English catch are the increase in the proportion of cod and the decrease in herring. The proportion of cod rose from the start of the century to the Second World War with the expansion of the distant-water trawl fishery on the richer Arctic grounds where cod dominate the catch, and which provided progressively more fresh fish for urban markets. The run-down of the herring catch is associated mainly with changing market and trade conditions—particularly the inter-war collapse of the markets for cured herring in Eastern Europe. The decrease of the haddock catch is largely associated with intensifying competition and over-fishing on North Sea grounds; and decreases in hake, which are caught mainly off the west coast, are associated also with intensifying competition and with the run-down of the trawler fleets at the West Coast ports of Fleetwood and Milford Haven. The decrease of plaice in the inter-war years was also due to over-fishing of the main grounds in the southern North Sea, and the subsequent recovery is largely due to supplies of Arctic plaice coming from distant-water trawlers.

In Denmark before the Second World War the fisheries were mainly for national food requirements, and the most important species in the catch were cod and plaice, by both weight and value. Although landings of both of these have more than doubled in the post-war period, they now form a small fraction of the catch, which has been greatly expanded by the rise of the reduction fisheries of herring, sand-eels, Norway pout and whiting.

Norwegian figures show that cod and herring have always been the most important species in landings, but in the post-war period there has been a considerable widening of the stocks exploited, to the point that catches of cod and herring together comprise less than half the total. Up to the Second World War, herring (with sprat) accounted for from 60 per cent. to 75 per cent. of all landings, and cod from 15 per cent. to 30 per cent., and the two vied with each other for the greatest value. The expansion of reduction fisheries for capelin and mackerel have been the main change in Norway, although there has also been expansion of landings in food fish like haddock and halibut. In recent years herring have been around 40 per cent. of all landings, and cod less than 10 per cent.

REFERENCES

1 M. M. Ovchynnyk, 'Development of Some Marine and Inland Russian Fisheries and Fisheries Utilisation', in *Atlantic Ocean Fisheries*, ed. G. Borgstrom and A. J. Heighway, 1961, p. 268.

2 G. Borgstrom, 'The Atlantic Ocean Fisheries of the U.S.S.R.', in *Atlantic Ocean Fisheries*, p. 265.

3 N. P. Sisoev, *Ekonomika Ribnoi Promyshlennosti*, 1966, p. 281.

4 H. Wood, 'Fisheries in the United Kingdom', in *Sea Fisheries. Their Investigation in the United Kingdom*, ed. M. Graham, 1956, p. 40.

5 N. P. Sisoev, *Ekonomika Ribnoi Promyshlennosti*, 1966, p. 80.

6 E. D. Kustov, *Geografiya Ribnoi Promyshlennosti*, 1968, p. 7.

CHAPTER VI

Fleets and Fishermen

In fishing operations, matters of capital and labour are very largely those of fleets and fishermen. With the modern advances in technology and wage-levels, the trend is for fisheries to become more capital-intensive and less labour-intensive—the common trend in production throughout the developed world. Throughout Europe considerable variations exist in sizes of fleets and labour forces, and in the trends of investment in both.

THE FLEETS & THEIR EQUIPMENT

Vessels and their gear[1] are the capital equipment employed in the catching sector of fisheries. They show variations in numbers and type which even now are due in part to long tradition, although there are marked trends towards greater uniformity. Data on capital inputs to fleets are rare and on the European continent comparison can be made only of numbers, tonnages and power of vessels. It is possible, however, to examine the factors that determine the scale of capital investment. As a general rule both absolute and proportional financial returns increase with the scale of investment, although exceptions to this are not difficult to find.

In the smaller-scale fisheries, which are generally labour-intensive, very varied boat and gear types are employed, although usually their cost is relatively low. Small traditional craft are open boats everywhere and are essentially short-range inshore craft. In the warmer conditions of southern Europe even relatively large craft may have little investment in decking or in crews quarters, while in the North Sea virtually all craft over 30 feet are fully decked. At the simpler level of equipment is a series of types of traps, lines and nets: the last-named are usually the most important, and include a variety of trawl, ground-seine, ring-net, set- and drift-nets, all capable of being hauled by hand. These simpler types of equipment are most commonly used in Mediterranean countries, but are also employed in inland waters, and to a more limited extent in the seas off north-west Europe. In the more advanced countries, these simpler gears tend to be used for specialized purposes, and include creels for lobsters and trap nets for salmon; but the Lofoten cod fishery is an example of a long-established labour-intensive fishery which has survived, although partly as a result of legislative protection. Even at the simplest levels of operation there is now a strong movement towards at least the installation of engines in the boats employed; this is an obvious necessary first step in promoting greater productivity. The craft used for the labour-intensive fisheries have one advantage over bigger boats in their greater versatility, and they can generally be easily adapted to a variety of uses if desired.

The earliest engines installed on fishing craft were in the 1880s, when they were of course steam-powered and coal-fired; and the space required for boiler and bunkers as well as for the engine itself restricted it to the largest vessels then operating. Steam vessels were developed especially in Britain and Germany, and to a considerably less extent in such countries as the Netherlands, France and Norway. From the earlier years of this century, fishing craft also had motor engines installed, and these were more compact and could be used for small vessels also. This led to most of the catches in north-west Europe being made by powered craft in the inter-war period, and post-war years have seen the adoption of diesel power as the most economic for the great majority of fishing vessels.

From small open craft there is a range in vessel size up to the biggest factory-trawlers and factory ships, which may be over 10,000 tons. Vessels of up to 70 to 100 feet are usually of wooden construction, but over this size higher levels of investment are usually required for steel vessels. The levels of investment vary from under £5,000 for small decked craft to over £1 million for factory trawlers, and the mother ships constructed for the U.S.S.R. must cost several times this figure. A main aim with larger vessels is to permit operation at long range, and this involves greater carrying as well as catching capacity, together with the use of more space for fuel and stores, as well as better crew accommodation. It often now involves the installation of freezing equipment also.

On vessels above about 30 feet, the marked preference is now for types of gear which can be hauled by winch or capstan. While there are elements of hand-hauling in some of the long-lines, ring-drift- and gill-nets which are employed, it is usually possible to perform at least part of the hauling by winch, although these types of gear are still relatively labour-intensive. What is now a rare example of big labour-intensive units are the Portuguese dory-schooners, working the Grand Banks; they are usually about 120 feet long, and function as mother-ships for the one-man dories from which cod are caught by lines. On these vessels the crew is about seventy, compared with perhaps a dozen for a trawler of the same size. The Faroese line-fishery for cod from shore bases in Greenland, which is also operated from small open craft, is another example of a labour-intensive fishery.

The most common gears employed by bigger craft are now trawls and purse-seines: these can be made to the biggest specifications, and a limited amount of manual operation is needed to shoot and haul them. (Some hand operation is required on side-trawlers, but the modern trend is towards stern-trawlers where the work can be more fully mechanised). Trawls are used mainly for taking demersal fish on the bottom, while the purse-seine operates by ringing shoals of pelagic fish. Mid-water trawls, usually towed by pairs of vessels however, now take a considerable share of pelagic fish in northern Europe, although it is rare for demersal fish to shoal so thickly and at suitable depths for catching by purse-seine. Trawling in the earlier part of this century was largely the preserve of Britain and Germany, but the East European countries now have big trawler fleets, and all countries of Western Europe are now involved, with Spain having one of the largest fleets. Purse-seining, in its modern form by which it can be operated in the open sea and hauled by a power block, is a modern innovation from the latter 1950s, and is especially deployed in Iceland, and Norway, but has also been adopted by the U.S.S.R. and other nations. Big vessels of over 100 feet are also employed for long-lining, especially in Norway and the Faroe Islands, although this is labour-intensive, especially for the task of baiting. There is a trend now to incur the additional investment of heated shelter-decks for the long-range vessels operating in the Arctic, where most of the work can be done in more comfort and safety.

Ground-seining is a method which involves the use of a bag-net like the trawl, but which is operated with ropes rather than steel warps. It is generally employed by smaller and less powerful vessels than the trawler, but has the advantage that it can sweep a bigger area of ground in a given time. It is used particularly in Denmark and Scotland, although there is some tendency now for even smaller vessels to instal bigger engines and employ the trawl.

Materials for gear have seen important innovations with the employment of nets, lines and ropes made of nylon and similar fibres. Although necessitating additional initial investment, such materials are rot-proof, and much longer-lasting than the older gears of hemp, cotton and manilla. The new materials are now near-universal in north-west Europe. Considerable effort also has gone into the improvement of winches and capstans, and these are now often hydraulically operated on bigger vessels.

In the post-war period investment in various fishing aids has much increased. Virtually every vessel over 30 feet now has an echo-sounder, which shows water depth and pelagic shoals below the boat; and more elaborate instruments can show demersal fish on the sea bed, while sonar instruments can detect fish shoals in the sea for a distance of several hundred yards around the vessel. Precision navigation and position finding equipment and radio transmitters are now in general use by even inshore fishing craft, and radar is also becoming common in Northern Europe, where visibility is more apt to be restricted.

While the basic work of shooting and hauling gear is now substantially mechanized on the more advanced fishing fleets, there is a very limited amount of mechanized handling of catches once they are on board. This is partly because space is at a premium even aboard relatively large boats, but also because of the difficulty of getting machinery which is sufficiently versatile, mobile and dependable. Only on the biggest ships is there much machine-gutting, and only on factory-trawlers and mother-ships is there much use of conveyors in moving the catch. Stowage is still very largely by hand on vessels of

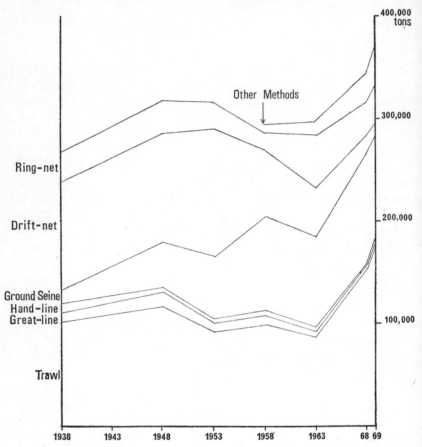

Figure 11. Trends in Scottish landings by method of capture.

all sizes, although it is common now for fish to be shelved or boxed, rather than stowed in bulk; this makes for easier eventual unloading.

The trend towards more capital-intensive fisheries is illustrated by Figure 11, which shows the landings in Scotland by method of capture from 1938 to 1969. During this period, the catch has become increasingly dominated by fish caught by trawl and ground-seine gear, which are very largely power-hauled, while the more labour-intensive methods of lining and drift-netting have progressively lost ground. The decline in the trawl catch and the increase in the ground-seine until the early 1960s is largely due to a higher measure of government support being given to the vessels of the seine-net fleet, which are smaller; and the subsequent steep rise in the trawl catch is due to many of these smaller vessels installing more powerful engines and using trawls for herring as well as white fish. The recent growth in 'other methods' is due to a section of the fleet adopting the purse-seine and the nephrops trawl. There has been a fairly constant proportion of the catch taken by the ring-net, which (while relatively capital-intensive) is used by small craft only; bigger vessels now employ the purse-seine (effectively an enlarged ring-net), for which more power-hauling is possible.

In all, there is a general tendency throughout Europe for the effective catching power to be concentrated in a smaller number of vessels; and the length, tonnage and engine-power of fleets are all tending to increase, while both gear and fishing aids become increasingly sophisticated. This demands an accelerated rate of investment, and as the fishing industries of most countries are economically weak, most European governments now have instituted schemes for aiding investment.[2] In Iceland any aids to investment are financed from within the industry, and in several countries aid is limited to organized lending at market rates: this applies in Denmark, Portugal and Greece. In most cases, however, financial aid in the form of low-interest loans or investment grants is given, and for the British inshore fleet and the Italian south, direct investment subsidies of 40 per cent. or more may be given. In the post-war period, investment in the fishing industry has certainly been greatest in Eastern Europe, particularly in the U.S.S.R. While the criteria for the assessment of the return on this investment are not clear, a substantial number of the vessels have been built in Western European yards, which suggests that the unit costs involved in building the modern fleets have been comparable with those incurred in Western Europe.

NUMBERS & SIZES OF VESSELS

Statistics are kept of the numbers of fishing vessels in most countries, but the great variety in methods of organizing and presenting the statistics renders impossible any other than crude comparisons of catching power between different countries and regions. It is very generally the case that while large numbers of craft are involved on some scale in fishing, the effective catching power is concentrated in small numbers of relatively large vessels. In some countries, however, most of the smaller, less important boats are excluded from published statistics, while in others all the small craft may be registered and recorded. There is also a considerable variation in the extent to which the catching power has become concentrated in large vessels.

In Europe, the number of fishing craft is almost certainly largest in the U.S.S.R. The number in the U.S.S.R. in 1961 was stated as 100,000, 23,000 consisting of modern steam or diesel vessels; and, in 1966, the European coasts of the U.S.S.R. had 50 per cent. of the national total of powered craft, and 62 per cent. of the total horse-power installed in fishing vessels.[3] Italy, with 45,760 registered vessels in 1965 is next in vessel numbers, but only 17,282 were motor-driven, and about 60 per cent. of the total were under 3 tons. Norway, with 39,895 boats in 1966, also has a large fleet: 29,968 of these were open boats, and 32,755 were less than 30 feet, while only 400 were over 100 feet and only 511 steel built. Of the Danish fleet of 12,258 in 1968, two-thirds were power driven, but only 310 were over 50 tons. Although Portugal has a considerable distant-water fleet, the large number of small craft with engines is (like Italy) typical of the Mediterranean zone: in 1966, of a fleet of 10,759, there were 7,642 sail and row-boats, and 8,840 were under 5 tons, while only 268 were over 50 tons.

For the countries in which the smaller craft are not included in the published figures, more meaningful comparisons of average tonnage can be made. Of these Poland, with her big post-war programme of construction of distant-water vessels, has the greatest average: in 1965 the tonnage had doubled since 1960 and there were 1,374 ships with an average tonnage of 105. The average for the Icelandic fleet of 654 was 94 tons, and for the 383 Belgian vessels 78. Although West Germany and England have big distant-water fleets, the numbers of smaller vessels brought down the average tonnages to 65 and 49 respectively in 1965.

TRENDS IN FLEET SIZES

Trends in numbers of vessels over a period can be seen, but this only in part reflects changes in catching power. In Norway the number of unpowered open boats appears to have remained substantially constant at around 50,000 from the early years of the century until at least the late 'fifties, although the installation of engines in the really effective part of the fleet was substantially complete by the inter-war years. In 1946 there were 18,711 open craft with engines, and 12,252 decked boats. The most noteable change in the post-war period has been the reduction in numbers of smaller decked boats in the 30 to 59 feet class: this group decreased from 9,109 in 1955 to 5,804 in 1966. During the same period the numbers in the over 60 feet class remained constant, between 1,300 and 1,400, although its average size and catching power greatly increased.

In Scotland, the increase in catching power in the present century has been achieved very largely through the run-down of the labour-intensive drift-net fishery (which dominated activity) in favour of other methods, especially seining and trawling. There has been little change in average tonnage since 1913, although the number of vessels in 1965 stood at 2,937 and was less than one-third of the 1913 total. In England, with the early development of steam-trawling, there was a big increase in average tonnage till the First World War, and in 1913 there were 9,415 vessels of which the average tonnage was 21·5; thereafter there has been a progressive run-down of the fleet with a slow rise in average tonnage, and in 1965 the 4,987 registered vessels had an average tonnage of 49.

For the post-war period, the fullest comparative statistics are those for the European Economic Community. Figure 12 shows numbers of powered vessels and gross tonnages for the years 1953 and 1965 for the five countries with marine fisheries.

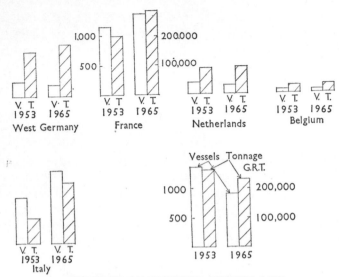

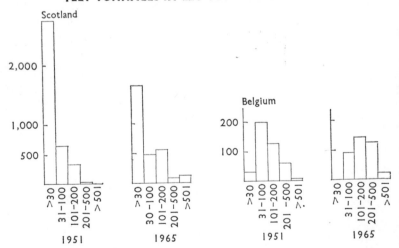

**NUMBERS OF POWERED VESSELS AND
FEET TONNAGES IN EEC COUNTRIES 1953 and 1965**

**NUMBERS OF VESSELS IN DIFFERENT POWER CLASSES
IN SCOTLAND AND BELGIUM 1951 and 1965**

Figure 12. Trends in tonnages and horse-power of fleets in sample countries.

An increase in tonnage was general during this period, even for West Germany, the Netherlands and Belgium in which the numbers of powered vessels have declined. The increase in numbers of powered craft in Italy and France is not inconsistent with the trend towards smaller fleets, as these indicate the installation of engines in boats previously powered by oar and sail. The higher tonnages in West Germany, Belgium and the Netherlands are evident, and this illustrates their greater concentration on capital-intensive fisheries. The average vessel tonnage in both the Netherlands and West Germany increased by nearly 50 per cent. between 1953 and 1965. For the German trawler fleet, which has increasingly dominated the nations' fisheries, a longer sequence of statistics is available. This shows the increase in average tonnage from 40 in the 1880s to 150 in 1953,[4] and in 1968 to over 950 tons.

The post-war trends in installation of more powerful engines are also illustrated in Figure 12 by the composition of the fleets in Scotland and Belgium according to power-classes for 1951 and 1965. In Scotland there was a reduction of 40 per cent. in the under 30 h.p. class, and an increase of 70 per cent. in the 101–200 h.p. class; and while no vessels were of over 500 h.p. in 1951, there were 103 in 1965. In Belgium the under 30 h.p. class was eliminated in this period, while that of over 200 h.p. doubled.

EMPLOYMENT IN FISHING

The distribution of employment in fishing in Europe is primarily governed by the scale of the industry in different countries and regions; but it is also very much conditioned by the degree to which fisheries have become capital-intensive. The continent has developed to such an extent that fishing is now seldom a dominant employer at even a regional level; the secondary and tertiary economic sectors now occupy the major part of the labour force, and where primary employment is still the main sector the majority are employed in agriculture rather than fishing. The highest levels of employment are found in the U.S.S.R., and in the Mediterranean area: particularly in the latter there is a big primary sector, and fisheries are often still labour-intensive. On the other hand, in most of the main fishing nations of the western

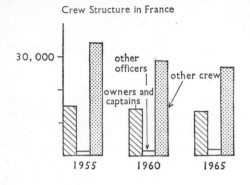

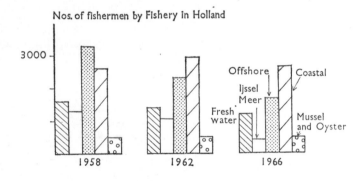

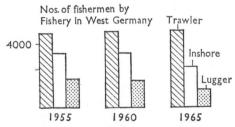

Figure 13. Changing structure of fishing labour force in France, West Germany and Netherlands.

seaboard, employment levels are relatively low, and fisheries more capital-intensive.

Of the leading fishing nations of the world, no precise

employment statistics are available for Japan, China and
Peru. It is certain that the numbers of fishermen in the first
two are well in excess of those of any European country, but
the recently developed Peruvian fisheries are capital-intensive,
and have been estimated to have a labour force of only 20,000.[5]
It has been estimated that Japan has around one million
fishermen. Fishing employment in the U.S.S.R. is stated
to be around 170,000 and the majority of these are in Europe:
this includes a substantial proportion in inland fisheries,
and in those of the Caspian and Black Seas. Spain, with
106,000 men recorded on powered craft in 1961,[6] is the only
country in Europe in which employment is on the same
scale as in the U.S.S.R.: if allowance is made for men in
row- and sail-boats, and in inland fisheries, the Spanish
totals are probably above those of the European part of the
U.S.S.R. Italy, Norway, France and Portugal have fishing
labour forces in the range between 35,000 and 55,000; among
the main fishing nations Britain has c. 22,000, Denmark
c. 16,500 and only in Sweden, Poland and Greece of the
remaining nations are the figures in excess of 10,000. Despite
the scale of their landings, the numbers of fishermen in West
Germany and Iceland are well below 10,000, and they have
the most capital-intensive fisheries of Europe.

In a limited number of instances, details of employment
are available by sectors of the fishing industry. In England,
the trawling sector has had over half the employment since
before the First World War, and since the early 1950s has had
over 70 per cent. In West Germany, trawling is also dominant,
with over half the total labour force, while the middle and
coastal water fishery has 30 per cent. and the (now shrunken)
lugger herring fishery less than 10 per cent. Dutch statistics
have been published for both sea and inland fisheries (Figure
13): in 1966 the major sector was the 'coastal and cutter'
sector with 42 per cent., while deep-sea men accounted for
26 per cent. 19 per cent. were engaged in fishing inland
fresh water, and 7 per cent. were specialized crab and mussel
fishermen, while 6 per cent. were engaged in exploiting the
Ijssel Meer: with the construction of new polders, this last
sector has been contracting rapidly and its numbers were
less than one-third of the 1958 figure, when it accounted for

14 per cent. of the labour force. In Portugal, sixteen employment groups are recognized, not including the north-west
Atlantic cod fishery; of these the main ones in 1966 were line
fishing (with 32 per cent.) and trawling (with 29 per cent.).
Poland shows a distinction between state enterprises, which
are mainly long-range enterprises with large vessels (particularly trawlers), fishermen's co-operatives and independent
men: in 1961 the three sectors had 60 per cent., 15 per cent.,
and 25 per cent. of the employment respectively.[7] In East
Germany, however, the 2,000 independent fishermen in 1959
constituted the major part of the work-force.[8]

Within individual countries, there are big variations in the
distribution of fishermen. In Sweden, more than 60 per cent.
of the full-time fishermen are resident on the relatively short
west coast which has immediate access to the North Sea
grounds; and on the Baltic coast, which is three times as long
but which depends on the poorer resources of the Baltic,
there are under 40 per cent. In Norway, fishing has always
been relatively most important in the north, and the three
northern counties out of the national total of eighteen have
nearly 50 per cent. of the fishermen. In Scotland, the catching
power and labour force are concentrated in the north-east,
and here are to be found a full 50 per cent. of the fishermen,
on about one-tenth of the total length of coast. Italy has its
greatest numbers of fishermen in the south, and about 30 per
cent. of them are on Sicily alone. While fishing is important
along almost all the Portuguese coast and in the islands of
Madeira and Azores, of the four recognized divisions the
northern area (from Figueira da Foz to the Spanish border)
has a higher level of employment in fishing, with 36 per cent.
of the total.

TRENDS IN NUMBERS OF FISHERMEN

It has been suggested that there has been a widespread
decline in the labour-force engaged in fishing in Europe
since around 1900. This may be the case, with the general
run-down in employment in the primary sector and with the
growth of capital-intensive enterprises in fisheries. Published
statistics, however, appear to indicate this with certainty

only for Britain, although a general decline in employment is demonstrable in most countries since the Second World

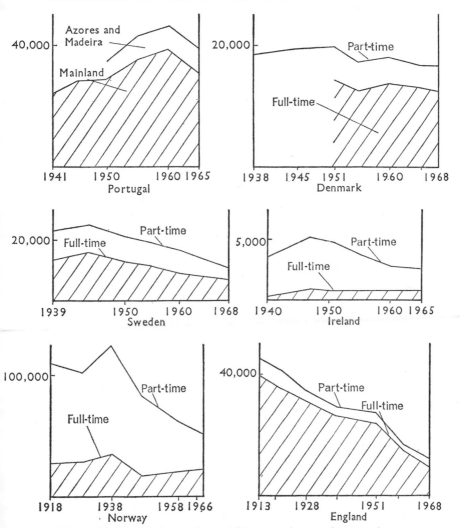

Figure 14. Trends in numbers of fishermen in sample countries.

War (Figure 14). Even here, however, there are exceptions, with the rise in fishing man-power in the U.S.S.R. and Poland, and the fluctuations in both Italy and Portugal in the period since 1950. For the earlier period, the numbers in Iceland rose until 1920, since when there has been a fairly gradual decline; and the Norwegian fishermen increased in numbers till the early 1930s, although since then they have decreased to less than 40 per cent. of the maximum. The long-term run-down, however, has been greatest in Britain, as a result both of a move towards more capital-intensive enterprises, and of the big contraction of the formerly dominant herring fishery with changed market conditions from the First World War. The number of British fishermen before 1913 exceeded 84,000, but by 1968 was down to under 22,000, a decrease of 74 per cent.

The decrease since the Second World War has proceeded at varied rates. Recorded fishing employment in Italy, despite fluctuations since 1955, has decreased by over 50 per cent. since 1946, and in Sweden by a comparable amount since 1944. The post-war decline in Scotland also approaches this proportion, while the main decrease in England has been post-war, and the labour force in 1968 stood at only 40 per cent. of the 1938 figure. In Spain, France and West Germany the post-war decrease has been by over one-third in each case. Of the West European nations, the decrease has been least in Denmark with her vigorous post-war expansion in fisheries; indeed manpower expanded somewhat in the immediate post-war years, and the subsequent decline has been to 86 per cent. of the 1945 figure. Developments in Poland and the U.S.S.R. are in strong contrast to those in Western Europe. In Poland, recorded numbers increased from 1,800 in 1938 to 5,218 in 1955 and to 10,655 in 1965; the latter figure representing a six-fold expansion over 1938. A big incentive here has been the much higher earnings the men, largely recruited from families not traditionally involved in fishing, could obtain.[9] A survey about 1960 revealed that they obtained on the average two-and-one-half times the earnings they could get in shore employment. Earnings incentives have also featured in the U.S.S.R. in the build-up of fisheries and personnel aboard vessels involved in the industry increased ten-fold between 1940 and 1959.[10] In the U.S.S.R., rates of pay for fishermen working in the Arctic are double the basic rate, and for those in the North Atlantic 1·8 times the basic rate.[11]

Part-time fishermen are still to be found in all countries. Although the modern trend towards specialized full-time occupations has reduced the proportion of them in nearly all countries, they still constitute a relatively large proportion of the labour force in many. It appears this will continue into the future, as shorter working weeks in other occupations are giving men resident at the coasts an opportunity to increase their incomes by fishing. On the other hand, the numbers of fishermen with small holdings, who in pre-industrial times probably accounted for the major part of the labour force, are going down more steeply.

Even in a country which has developed as fast as Sweden, the proportion of part-time fishermen has declined only from 40 per cent. in 1939 to 35 per cent. in 1968, although the fall in absolute numbers has been from 9,671 to 3,551. In Denmark the proportion has remained fairly constant during the post-war period at around 25 per cent., but in Norway changes have been greater with the big decline of the joint occupation of smallholder-fishermen: part-time fishermen accounted for 79 per cent. of the labour-force in 1948, but 55 per cent. in 1966. A similar trend is instanced in Scotland, where crofter-fishermen were 30 per cent. of all fishermen in 1920 but 8 per cent. in 1966, when they were rather less than half the part-time labour-force. Ireland, where fishing was formerly very much of a secondary occupation for smallholders, appears to have shown one of the greatest proportional changes: the part-time sector of the labour force was 90 per cent. in 1940, but 70 per cent. in 1965. While it is not possible to show a trend in the U.S.S.R., it was stated in 1959 that one-third of the catch was taken by 2,000 kolkhozes in which fishing was prosecuted along with farming and other employments.[12] However, this does not necessarily indicate a big number of part-time fishermen, as there is much specialization of labour within the kolkhozes.

Within the work of fishing itself, changes have taken place in the employment structure. The tendency for more of the fishermen to be employed on bigger vessels is shown by the changing structure of the West German labour force shown in Figure 13. Over the decade from 1955 to 1965, there was a slight rise in numbers in the crews of deep-sea trawlers, but there was a 27 per cent. fall in inshore fishermen, and a 35 per cent. of those on herring luggers. The changing structure of the French labour-force in Figure 13 shows a 27 per cent. decrease in deck-hands, and a 7 per cent. decrease in the group of skippers and owners, while 'other officers' rose by 16 per cent.: this reflects a contracting fleet which is becoming more capital-intensive, and which accordingly needs fewer deck-hands but more specialized officers.

REFERENCES

1 For a description of different vessels and gear, see R. Morgan, *World Sea Fisheries*, 1956, pp. 65–119.

2 O.E.E.C., *Financial Support to the Fishing Industry*, 1965, pp. 21–39.

3 N. P. Sisoev, *Ekonomika Ribnoi Promyshlennosti*, 1966, p. 196.

4 T. Laevastu, 'Natural Bases of Fisheries in the Atlantic Ocean', in *Atlantic Ocean Fisheries*, ed. G. Borgstrom and A. J. Heighway, 1961, p. 30.

5 *The Times*, Aug. 28th, 1970.

6 *Economic and Social Development Program for Spain 1964–67*, prepared by the Commissioner for Economic and Social Development Planning, Presidency of the Government of Spain, 1965, p. 124.

7 L. G. B. Butcher, 'Looking at the Fisherman', in *Some Aspects of Fishery Economics*, ed. G. M. Gerhardson 1964 II, p. 25.

8 G. Borgstrom, 'The Atlantic Fisheries of the USSR', in *Atlantic Ocean Fisheries*, p. 313.

9 L. G. B. Butcher, *Ibid.*, p. 26.

10 M. M. Ovchynnyck, 'Development of Some Marine and Inland Russian Fisheries and Fish Utilisation', in *Atlantic Ocean Fisheries*, p. 275.

11 N. P. Sisoev, *Ibid.*, p. 300.

12 G. Borgstrom, *Ibid.*, p. 283.

CHAPTER VII

Productivity of Fleets and Fishermen

Most assessments of productivity in fisheries can only be crude, but the differences between the various parts of Europe are such that certain broad patterns can be discovered even with the incomplete data which are available. Within the field of productivity, criteria for assessment may be both volume and value of catches; and productivity of grounds, fleets and fishermen are all of importance.

General levels of productivity in the major sea areas

TABLE 1

CATCHES PER FISHERMAN AND PER VESSEL TON
FOR 1955 AND 1965 IN SELECTED COUNTRIES

Country	Catch/Fisherman				Catch/vessel ton (G.R.T.)			
	metric tons		$		metric tons		$	
	1955	1965	1955	1965	1955	1965	1955	1965
Belgium	35·0	31·8	5100	8010	3·05	2·00	445	509
Denmark	25·0	34·0	2090	5500	6·35	8·75	542	981
France	11·8	18·1	3520	5600	3·03	2·68	730	824
Germany (W)	85·5	77·1	6960	10900	4·96	3·37	402	474
Iceland	n.a.	n.a.	n.a.	n.a.	8·97	14·90	n.a.	790
Ireland	3·1	6·7	470	980	n.a.	n.a.	n.a.	n.a.
Italy	5·2	5·7	n.a.	2900	n.a.	n.a.	n.a.	n.a.
Netherlands	n.a.	55·4	n.a.	7750	6·06	4·07	386	569
Norway	20·1	47·5	943	3200	n.a.	6·51	n.a.	437
Poland	24·3	27·8	n.a.	n.a.	n.a.	2·03	n.a.	n.a.
Portugal	10·5	14·3	1220	1640	3·84	3·58	393	435
United Kingdom	37·0	40·1	4460	6950	3·23	3·49	392	589

n.a.: not available.

recognized by international conventions have already been discussed (Chapter V); it is also essential, however, to relate the yield of these sea areas to the effort expended in obtaining it.

Productivity of fleets would ideally be measured by the percentage return of their capital value; this is seldom possible,
although some examples are indicated below from Norway,
Scotland and the Faroe Islands. For wider comparative
purposes, it is best to relate the catches to the tonnage of vessels
which produce them. Capital investment increases with the
size of vessels, and although the rate of increase tapers somewhat in the bigger vessel ranges, it appears to be roughly in
proportion to vessel tonnage, to judge from the example of
Britain. Productivity of fishermen can be more readily
assessed, although a complication here is that there is nearly
always a part-time sector in the labour force.

Table 1 shows catches per vessel ton and per fisherman by
weight and value for twelve nations for which statistics are
available. The units employed are the metric tons and
American dollars of the F.A.O. Fisheries Year-Books, and the
period 1955 to 1965 shows something of modern trends.

CATCHING RATES
FOR DIFFERENT GROUNDS & FLEETS

Productivity of different fleets and grounds can be expressed
in a variety of ways: these include catch per haul, per landing
and per days' absence from port, but most useful for comparison
are the catches per unit of fishing time. For all sizes of vessels
and for all grounds, variations in catching rates show a great
amplitude. For individual vessels in demersal fishing the catch
may vary from zero up to (for large trawlers) several tons
per hour; and in pelagic fisheries the catch can vary from
several hundred tons per hour with big purse-seiners and
trawlers down to zero with all vessel types. There is a limited
amount of comparative data available for pelagic fisheries to
show the returns per unit of effort, and in any case the great
natural fluctuations in the stocks and their more erratic
locations render the interpretation of statistical data especially
complex. The most complete and best comparative data are
available for bottom trawling, and this is chiefly discussed here.

Comparison of different sea areas in their yield of demersal
fish can be made from the statistics published for the Norwegian,

German and British trawl fisheries: for the latter, the statistics are presented separately for the English and Scottish fleets.

In the British statistics, there are separate records for near- and middle-water fleets on the one hand, and distant-water on the other, and the two are not directly comparable, as the average vessel size and power is considerably greater for the distant-water fleet. There is a considerable range in catching power in the near- and middle-water fleets with vessels ranging from 70 to 140 feet. The English near- and middle-water vessels work all over the North Sea, and there are also important grounds off North-West Scotland. In the North Sea, the general level of international exploitation decreases from south to north, while catch levels increase. Between 1956 and 1968 in the southern North Sea, the average annual yield per 100 hours' trawling ranged from 77 to 188 cwt.; in the same period in the central division the variation was from 116 to 170 cwt., and in the north it was from 447 to 827 cwt. The latter area was worked by middle-water trawlers only, as were the grounds off North-West Scotland, where the range per 100 hours' fishing was from 275 to 615 cwt. The level of yield from Faroe grounds exploited by English middle-water trawlers was of the same size order as that from the northern North Sea at 418 to 615 cwt. The Scottish trawler fleet shows a comparable pattern of yields for all but the southern North Sea grounds, which they do not work to any extent: the main difference is that there is less difference between rates of catch in the central and northern North Sea areas, as both near- and middle-water trawlers operate in both. Additional details on the catching rates of British trawlers in the North Sea have been given for the inter-war period by Sir A. Hardy: detailed maps for the recognized rectangles (i.e. divisions of $\frac{1}{2}$° latitude x 1° longitude) show the greater yields of both cod and haddock in the north, and catches of over 25 cwt. per 100 hours' fishing came exclusively from north of lat. 54°N; and within this area, rectangles yielding over 50 cwt. per 100 hours varied seasonally.[1]

For the English fleet operating on distant-water grounds, the catches over the post-war period have been consistently over 1,000 cwt per 100 hours' fishing, apart from Iceland where some years have been below this figure. There was, however, a general decline in catching rates from the mid-fifties to the mid-sixties, since when there has been some improvement largely as a result of the elimination of older, less profitable vessels. In the mid-fifties, Spitzbergen, the Norwegian Sea and Bear Island grounds were all yielding average catches over the year of around 2,000 cwt. per 100 hours' fishing, and Iceland around 1,500 cwt. The subsequent decline in the rate has varied from 40 per cent. to 50 per cent.

The West German and Norwegian deep sea trawler fleets also show a generally higher rate of catch on more distant grounds, although the main difference here is between North-East and North-West Atlantic areas. In recent years West German catches have been around 12 metric tons per fishing day at Iceland; yields from the Norwegian Coast tend to run some 20 per cent. above this, but those from West Greenland and Labrador have been about double the level at Iceland, while those at Faroe have been around 25 per cent. lower. Norwegian deep sea trawlers' yields from the Norwegian Sea have been fairly consistent in the 1964–68 period, varying from 0·43 to 0·58 metric tons per hour fishing; while Barents Sea and Bear Island yields have reached this level, in some years they have been only half. The greatest yields, however, come from the North-West Atlantic grounds of West Greenland, Newfoundland and Labrador, where the general level has been about twice that of the Norwegian Sea, and has reached 1·22 metric tons per hour. The all-over average catching rate of the Norwegian fleet has markedly improved in recent years, increasing from 0·41 metric tons per hour in 1963 to 0·67 in 1968, or by 63 per cent.; and with the longer working days of more modern vessels, the increase in the daily catch rate was 75 per cent. Another instance of the greater yields that justify the working of distant grounds is that of the Portuguese trawler fishery, although the usual size of vessel here is much smaller than the distant-water fleets already discussed. The general level of catching rate in Portuguese waters in post-war years has varied between 0·10 and 0·15 metric tons per hour, while on the coasts of Mauritania and Senegal it has fluctuated between 0·25 and 0·50 metric tons per hour; the general level of vessel size for the latter fisheries is over 150 tons, while the big majority of vessels operating on the Portuguese coast are below this size.

While no average catching rates are published for Russian fleets, their large B.M.R.T. vessels (factory-freezer stern-trawlers of 2,500 or more tons) can catch up to 70 to 80 metric tons per day[2]; and the general rate for one fleet taking redfish at Newfoundland was 10 to 12 metric tons per hour.[3]

Lower catching rates are, of course, characteristic of smaller vessels, as is instanced by those of the Scottish seine-net fleet, of which the rates are in the range of 10 to 50 cwt. per day's absence, the higher yields coming generally from grounds further off-shore, which are exploited by bigger boats. In Denmark seiners tend to have rather higher catching rates than light trawlers, but the landings at Esbjerg show that the general level of catching for both is in the range of 100 to 200 kg. per haul.

PRODUCTIVITY OF FLEETS

Productivity per vessel ton varies less than productivity per man, indicating that although the range of variation in returns to capital is big, it is less than that to labour. It also shows that the catches of the more simply-equipped men are relatively good in relation to their investment in gear.

Proximity to the grounds exploited is an important factor in the catch per vessel ton. Iceland, the well-equipped modern fleet of which spends a minimum of time in steaming to and from the grounds, had the highest rate in 1965 at 14·90 metric tons/vessel ton, although in some more recent years this figure has been reduced by almost one-half in poor seasons; and Denmark, which also has the advantage of proximity to the grounds worked, was second to Iceland with a figure of 8·75. Norway, with most of her catch coming from nearer waters, was third with 6·51. Although fisheries are capital-intensive in West Germany, Britain and France, the catch weights per vessel ton were considerably lower. This is largely due to the time lost by vessels in making the passage to and from Arctic grounds, but also because part of their catch comes from their inshore fleets working grounds which are less rich. The 1965 Polish figure of 2·03 suggests that distance from grounds also depresses the catching rate of the modern fleets from Eastern Europe. Between 1957 and 1961 the catching rate in Spain varied between 2·44 and 2·76 per vessel ton,[4] and is in the same category as those of Britain, West Germany and France.

In many of the European nations, the catch weight per vessel ton has decreased in recent years, and this reflects the deteriorating returns to investment which have led to difficulties, and to increasing government investment aid in many cases. Over the 1955–65 decade, the catch weight per vessel ton decreased by some 30 per cent. in Belgium, West Germany and the Netherlands; for the deep-sea trawler fleet which dominates West German fishing, the rate decreased 40 per cent. There were also decreases in the rate in Portugal and France, although it improved in Iceland and Denmark, thanks to an expanding herring fishery, and in Britain to a rationalization and contraction of the fleet.

In the catch value per vessel ton, the ranking of the countries is considerably changed because of the different values of fish at first sales. The position of Denmark and France at the head of the list is due to the relatively high proportions of the catch which go fresh to human consumption, and in France the average value is enhanced by the relatively high proportion of shell-fish and molluscs. The high catch value in Iceland per vessel ton is due to the very big volume of landings rather than high prices. The low Portuguese value is related mainly to the internal market of limited purchasing power in a less fully developed country.

Trends in the nominal value of catch per vessel ton have been upwards in all countries in the post-war period, but inflation has largely cancelled these, or rendered them negative in real-value terms. The rises in the 1955–65 period in Denmark (81 per cent.), Britain (51 per cent.) and the Netherlands (48 per cent.), however, do represent increases in real value, and indicate improved efficiency in production. While figures of catch per vessel ton are not available for the U.S.S.R., the average catch for B.M.R.T. vessels—the main component of the distant-water fleets—rose from 5,070 to 7,300 metric tons between 1959 and 1965;[5] this represents an increase of 44 per cent.

PRODUCTIVITY OF LABOUR

The productivity of labour shows a high amplitude of variation, but has been generally rising as fishermen become better equipped. The highest values are shown by the main fishing nations of the Atlantic seaboard, and there is again some difference in relative importance according to volume and to value.

In the leading countries in 1965, average catching rates were in excess of 30 metric tons per man (Table 1). In West Germany, where catching is most dominated by big trawlers, the level was 77·1 metric tons per man; and the Netherlands, Norway and Britain all had over 40 metric tons per man. Although no precise figures for 1965 are available for Iceland, the rate for this favourably situated and well equipped country now is better than anywhere else in Europe and appears to be in the region of 150–200 metric tons per man. Countries in which small operators dominate the labour force have considerably lower figures than the leading nations, as is instanced by Ireland with 6·7 metric tons per man and Italy with 5·7. The countries which have been vigorously expanding their fisheries in the post-war period still have catching rates well behind the leaders: In 1960 the Spanish rate was c. 14·0 tons per man, that of the U.S.S.R. in 1962 c. 22·0 and of Poland in 1965 27·8 metric tons per man.

While the highest figures now are all to be identified with capital-intensive fisheries, it is significant that, at an earlier stage (in the 1940s), the Norwegian winter herring fishery, prosecuted largely with small craft and simple gear, could give around 40 metric tons per man for a three-month season. On the other hand, in the Lofoten cod fishery in the same country it has always been rare for the catch to reach 5 tons per man, even if it has been one of the richest demersal fisheries prosecuted by traditional methods.

The figures tabulated and discussed above take no account of the fact that a considerable part of the labour force in some countries is part-time. No figures are available for part-time fishermen in Italy, but its effective rate per man is certainly higher than that shown in the table, as fishing is often an ancillary occupation for small-holders. Most Irish fishermen

K

are still part-time, and if two part-time are reckoned the equivalent of one full-time man, the computed catching rate for 1965 rises by 54 per cent. to 10·3 metric tons per man. The Norwegian fisheries returns divide fishermen into 'full-time', 'main occupation' and 'secondary occupation': if the second of these is reckoned as two-thirds and the last as one-third, then the 1965 catching rate rises by 30 per cent. to 61·8 metric tons per man. Even in Britain, where capital-intensive fisheries with full-time men have developed over a longer period than in any other country, a correction made for part-time men raises the figure to 46·6 metric tons per man (i.e. by 16 per cent.). Such adjustments render more accurate the crude rates in Table 1, but for broad comparisons within Europe this clearly amounts to changes in detail rather than in essentials.

While general trends in catch weight per man were upward in the 1955–65 decade, it will be seen that in West Germany and Belgium they were falling: these are countries which were early relatively well equipped, and which have been meeting mounting competition from other nations. The greatest increases have been recorded by Denmark (36 per cent.), which has largely been through the rise of herring trawling; by France (53 per cent.), mainly through increased concentration on modernized middle- and distant-water fleets, and by Ireland (116 per cent.), which has been achieved by a marked trend away from part-time to full-time operation, and by the modernisation of the inshore fleet.

The longer-term trends in labour productivity by catch weight have been more spectacular. The rate per man in Britain has trebled since 1913, and has increased ten-fold in Norway since 1920, while Denmark's vigorous post-war expansion has led to a twenty-fold rise over 1938. The U.S.S.R. too has recorded an increase from 4·2 metric tons per man in 1932 to 10·2 in 1956[6]; and the 1962 figure of 22·0 appears to indicate a further doubling in six years. These improvements have been achieved not only in the large-scale capital-intensive state fisheries of the U.S.S.R., but also in the fishermen's co-operatives, which usually involve smaller-scale operation; in the co-operatives per capita production almost doubled in the 1950s.[7]

In catch value per fisherman, the leading countries are largely those in which fish is landed at auction markets and where most of the catch goes direct for food. West Germany is again prominent with U.S. $10,900 per man in 1965; and in the same year production on the highly capitalized trawler fleet was U.S. $13,000 per man. The relatively high values for Belgium, the Netherlands and Britain are notable; and although no precise data are available for Iceland, the landed value per man appears to rival that of West Germany in good years, despite the much lower unit value of landings. The lower unit values of Norwegian landings also have the effect of depressing the catch value per man, and the Spanish value of U.S. $3,250 per man for 1961 is of the same order of size, but was achieved with a smaller catch and a labour force twice as large. The low values in Portugal and Ireland are explained by the relatively labour-intensive nature of their fisheries; and although the Italian catch per man is less than the Irish in landed weight, it is more than three times the value because of the high proportion of more valuable species, and the landed value per man in Italy is of the same order as that in Norway.

Real increases in value of catch per man cannot be stated because of the effects of inflation; but the rise in nominal value is again prominent in Denmark, where in the 1955–65 decade the rise surpassed 115 per cent., and in Norway, where the rapid run-down in the labour force together with investment in more modern vessels led to a rise of 240 per cent. The more general rise, however, was in the range of 40 per cent. to 60 per cent., which scarcely compensated for the decreasing value of money.

INCOME AND EXPENDITURE OF FLEETS

(A) THE NORWEGIAN EXAMPLE

Norway is unique among the nations of Europe in the detail in which financial returns for her fisheries are published. These have been gathered systematically during the post-war years to show levels of returns and viability, which were previously obscure. They give details of returns to vessels and crews, and also of the various costs of operation for vessels of different size classes and in different fisheries. It is fortunate that there is a great range in the scale of enterprise and in types of fishery in Norway, and while no direct comparisons of viability are possible with other countries, the Norwegian example does at least demonstrate the relationship of particular variables to each other. The effects of environmental factors like resource abundance and distance can be seen in the statistics, and the year to year returns give ample evidence of seasonal variations. It is, however, only in part that data are regionally organized within Norway: some of the returns for vessels are broken down between north and south, and this latter section is occasionally sub-divided into eastern and western parts. A regional component is also included in some of the individual fisheries, like the Lofoten and Finnmark cod fisheries. Fishermen's incomes are published on a more detailed regional basis (by counties, or groups of counties), and are also shown by full- and part-time men, and by age groups. The form in which the data have been published has changed from time to time, and discussion is concentrated on the 1960s, particularly during the period 1963–67, when a series of comparable statistics was published.* This also shows a variety of situations as the fisheries recovered from a series of bad years around 1960. The returns for the 1960s also bear out many of the observations for the period up to 1960 made by Holm.[8]

As in general in the financial arrangements of fisheries, the proceeds from a Norwegian vessel are divided into three parts—those to the crew, the gear and the vessel itself, and the shares allotted to the latter two must cover both the running costs and capital depreciation in the enterprise. For the biggest vessels (those over 80 or 100 feet in length) these two are often grouped, as a consequence of their ownership being in the same hands. For the smaller craft, the crew share usually approaches, or even exceeds, one-half the proceeds, and indeed for the smallest vessels (under 20 feet) it is around 65 per cent. The bigger vessels represent more capital-intensive enterprises, and for them the crew share is a lower proportion,

---

* Throughout this period, the exchange rate of the Norwegian kroner varied between 19·96 and 20·02 to the £, i.e. it was almost exactly 1s. or 5p.

although even for the biggest trawlers and purse-seiners it is of the order of 35 per cent. to 40 per cent.

The division of the other part of the proceeds between vessel and gear depends largely on the relative value and rate of depreciation of the two. For drift and gill netting, gear and maintenance costs are high, and 25 per cent. to 30 per cent. of the receipts are allocated to the gear. The expenditure on purse-seines is also high and the gear share is in the 20 per cent. to 25 per cent. range, while in lining, trawling and ground-seining the gear share is around 15 per cent. In fisheries which employ the drift and gill nets and purse-seines, the share to the gear may be greater than that to the boat, although more commonly the boat share is two or three times that of the gear, and ranges from 15 per cent. to 45 per cent. of total proceeds for boats under 100 feet, while above this size the vessel and gear are usually reckoned together.

Since it is the very general practice—and indeed is reinforced by legislation—for fishermen to own all boats of under 110 feet, for many of them their income is that of a crew-man together with an addition proportional to their investment in boat and gear. The income of hired hands, however, is given by the crew-man share.

It is noteworthy that very few vessels average over 300 days (83 per cent. of the year) at sea; and the time actually spent in fishing is a good deal less, around 250 days (68 per cent of the year). For bigger vessels, the main reason for this is the time required to make the passage to and from fishing grounds; smaller vessels lose more time through engaging in seasonal fisheries and through bad weather. Maintenance work is also required on vessels of all sizes, although on smaller boats part of this is done by the crews themselves.

On the average, the total revenue rises rapidly with the size of vessel; but costs also rise steeply, and the average crew-man's earnings rise much less rapidly. The average figures for revenue all include a limited measure of direct government subsidy; the rate of subsidy varies for different fisheries and for different vessel sizes, and no statement is published of levels of revenue in the absence of subsidy. The three years 1961–63 have data published for vessels of all sizes. During these years the gross return varies from c. kr.8000 (£400) for small boats under 20 feet, to a range from kr.675,000 (£33,750) to kr. 850,000 (£42,500) for vessels over 120 feet. Big trawlers of over 300 g.r.t. are separately recorded, and for them the gross receipts varied between kr. 925,000 (£46,250) and kr. 1,120,000 (£56,000). Costs for vessels under 20 feet were as low as c. kr.2,300; for those over 120 feet they were from kr. 350,000 to kr. 400,000; and for big trawlers kr. 525,000 to kr. 600,000. The average crew man share was kr. 5,000 to kr. 5,250 for boats under 20 feet, kr. 17,600 to kr. 18,800 for those over 120 feet, and kr. 21,000 to kr. 23,000 for the big trawlers. The general level of returns rose for the fleet until 1966, although there was a fall-off (apart from the big trawlers) in 1967 and 1968. Vessels of over 120 feet had average gross receipts of over kr. 2,000,000 in 1966, and in 1968 those of stern trawlers of over 300 g.r.t. were in excess of that figure. Such ranges in returns show big variations in intensity of exploiting resources, and to some extent in mobility. The modern tendency is for the catching power to become markedly concentrated in fewer, larger vessels.

Average percentage returns on capital in the period 1963–67 have generally been in the range between zero and 30 per cent. for vessels of under 100 feet, and between zero and 50 per cent. for bigger sizes: this shows that all sizes of enterprise are prone to make losses in bad years, and the position of especially the smaller sections of the fleet in the absence of subsidies would be exceedingly precarious. On the other hand, the levels of returns in good years are much above the average for enterprises other than fisheries.

While such average returns indicate some of the economic variations and uncertainties involved in fisheries, they do not show the variations in performance between individual vessels. This varies greatly even within the same length and tonnage classes, and every year there is a proportion of vessels which make losses in almost every section of the fleet. This proportion can be above 20 per cent. especially among smaller craft, but even in the 100–119 feet class it has been as high as 30 per cent. in poor herring seasons. In the 1960s the gross returns from the 40–49 feet class, for example, have varied from under kr. 50,000 to over kr. 650,000, and those from the 80–99 feet class from under kr. 200,000 to over kr. 2,000,000; in both cases (as already indicated) the median value for returns was much closer to the lower figure. In net value of returns, the bigger vessels showed greater extremes; here the 100–119 feet class ranged from losses of over kr. 80,000 to profits of over kr. 520,000, while the returns from the 40–49 feet class generally varied from losses of under kr. 40,000 to profits of kr. 100,000 or even more. The share per man varied less than the gross and net proceeds, and although the average tended to rise with size of vessels, the share per man on the 40–49 feet class could on occasion reach kr.50,000, which was almost as good as on the best large boats, although for craft under 40 feet it was very rare to reach kr. 20,000. However a man's share on the smaller craft could be under kr. 4,000—or even kr. 2,000, although on bigger boats it was seldom under kr. 8,000.

Even when the averages for the different sections of the fleet are considered, profits characteristically vary greatly from year to year in the same fishery. This is due mainly to fluctuations in catch, but market conditions also play a part. Again it may be instanced in enterprises of varying scales. In the Lofoten cod fishery, boats of 40–49 feet operating with gill-nets had average net returns between 1963 and 1967 of from kr. 3,200 to kr. 10,200, with a mean of kr. 5,650; and the individual man share per week varied from kr. 230 to kr. 420, with a mean of kr. 325. Boats of the same size class working with long lines in the Finnmark spring cod fishery over the same period had averaged net receipts of from kr. 2,460 to kr. 11,700, with a mean of kr. 6,800; and the weekly share per man varied between kr. 305 and kr. 830, with a mean of kr. 580. Considerable though these variations are, they represent those from fisheries which are relatively consistent in their yield, and the same can be exemplified from the off-shore bank line-fishery with bigger vessels of the 60–79 feet group. Here the average net proceeds ranged from kr. 3,810 to kr. 29,800 with a mean of kr. 18,500, while the range of the weekly man share was from kr. 360 to kr. 740, with a mean of kr. 580.

By comparison, Norwegian experience with both ground-seining and bottom-trawling is that returns fluctuate more, and they have not been consistently profitable; but the most uncertain returns are characteristically met in pelagic fisheries. The biggest size of vessel engaged in the winter herring fishery for which net returns are consistently available is that of 60–79 feet. In this class average returns ranged from a net loss of kr. 1,000 to a profit of kr. 14,500, with a mean of kr. 4,100; the weekly man share varied from kr. 260 to kr. 690, with a mean of kr. 400. In the better seasons of the 1950s, the net returns for this class of boat were commonly over kr. 20,000.

There has been investment in Norway in the post-war period in big trawlers of sizes from 300 to 1,000 tons, and although there are operational advantages here in Norway's proximity to the grounds of the Norwegian and the Barents Seas, these vessels have the major disadvantage of not landing in a big competitive market. In the period from 1958 to 1966, these trawlers made an overall loss; although the average gross proceeds per vessel were around kr. 1,000,000 per year, the net annual return varied from losses of kr. 66,000 to profits of kr. 50,000. The stern-trawlers, which now comprise half the fleet, have done better in later years, and their net profits can approach kr. 450,000. The returns to the crews, at kr. 350 to kr. 720 per man between 1958 and 1966, were not conspicuously high compared with smaller vessels, but the weekly man share on stern trawlers has since been in the region of kr. 1,000.

Structure of Costs

The structure of costs in fishing enterprises nearly always shows fixed costs higher than running costs, and the main part of the former is composed of maintenance and depreciation. For the smaller section of the fleet in North Norway (under 60 feet), however, which is intensively operated, running costs exceed fixed costs. The structure of running costs varies with the type of fishery engaged in. Fuel costs are always a major component, and are related to the distance from base and the time spent in fishing. In trawling, fuel costs are particularly heavy because of the power required to pull a trawl over the sea-bed or through the water. For trawling and seining, fuel costs are usually over 90 per cent. of running costs, and in long-range trawlers may be 20 per cent. of total costs. In drift-netting, fuel costs are usually proportionately less but can reach the 90 per cent. level; but in most types of line-fishing bait is a bigger item, and for the off-shore fisheries is 40 per cent. to 60 per cent. of total operating costs. For longer-range fisheries, distance from base requires expenditure on ice or salt, although even in distant-water operation this does not reach 10 per cent. of costs.

Regional Variations in Profitability

Most of the statistics are presented separately for the three northern counties (Nordland, Troms and Finnmark)—'North Norway', and for the remainder of the country—'South Norway'. In the north, dependence on fisheries is particularly heavy. There is a consistent tendency for northern vessels to show superior levels of gross and net profit to those of the south, even though their working year is rather shorter: this is related to richer and more dependable fisheries resources, and is particularly marked with the smaller section of the fleet under 80 feet. These northern fisheries, however, tend to be more labour intensive (i.e. crews are larger), and the net profit of vessels per man-week can be higher in the south. Thus the net profit of the 40–49 feet class in 1964 was kr. 66,500 in the north and kr. 55,600 in the south, but the profits per man-week were kr. 270 and kr. 348 respectively. In the same year the net profit in the 50–59 feet class was almost twice as high in the north, but the profit per man-week was kr. 284 against kr. 333 in the south.

Significant regional variations in profitability can also be illustrated from the records of boats under 40 feet, for which in some years separate statistics are available for the Skagerrak Coast. In 1961 and 1962, the general level of net profit was considerably higher in North Norway than in the other west coast counties, and in many cases was double or better. The inshore fleet on the Skagerrak coast, however, has the advantage of proximity to the most heavily urbanized part of the country, and also spent 70 per cent. of its effort in trawling for prawns, which have a specially high market value. The levels of profitability were of the same order as those in the north, and for the 35–39 feet craft were considerably superior.

The difficulty that small, less mobile craft tend to encounter in extending their operations beyond traditional seasons is also illustrated. In the years 1961–63 there was generally a limited difference in the gross receipts for boats operating 120 to 240 days from those which worked 240 to 360 days, and indeed in more than half the cases the former had higher gross returns. Vessel profit per man-week also was greater in the great majority of cases, especially in North Norway, and could be almost 100 per cent. higher in extreme cases.

Fishermen's Incomes & their Regional Variations

Fishermen's incomes show some variation according to location, as well as to type of enterprise. The average income from fishing in the best counties is more than twice that for the lowest; but in all counties it is still usual for fishermen to get supplementary income from other sources, and this tends to be highest in those in which income from fishing itself is least. In the Tröndelag and along the Skagerrak coast, this can account for 30 per cent. of the average fisherman's income. Incomes are highest in areas which have invested most in big modern vessels, and these are mainly the counties of western Norway where most of the purse-seiners, distant-water line-boats and large trawlers are based. Here the average fisherman's income now runs above kr. 20,000 (£1,150) per year, and Möre og Romsdal is especially outstanding. Fishermen's incomes have risen rapidly in the 1960s; the rates of increase have been almost parallel for full-time, main-occupation and secondary-occupation men, but there have been widening differentials, and full-time men earn almost double the main occupation category. Over a third of all fishermen now earn over kr. 20,000 per year, and highest incomes are concentrated in the most active age-group of 25 to 45 years.

(B) SCOTTISH & FAROESE EXAMPLES OF FLEET COSTINGS

While not available in the range and detail of the Norwegian statistics, these also give perspectives on the profitability of vessel operation.

In the Scottish inshore fleet in the years 1961–65,[9] the

average gross earnings varied from a range from £3,750 to £7,300 for boats under 40 feet to one of £10,000 to £22,000 for the 60–80 feet class. The share to the crews varied from 40 per cent. to 55 per cent. and the labour share per man from £450 to £1,100, with some tendency to be higher on bigger vessels. Although percentage returns on capital ranged from 2 per cent. to 32 per cent., the proportion of boats making a loss in nearly all groups could reach one-third. In the absence of operating subsidies (which accounted for around 10 per cent. of the revenue for the fleet), even the sections of the fleet over 60 feet would have been in a difficult position and the sections under 60 feet would have made regular losses. However, since 1965, the efficiency of the fleet, stimulated by investment aid from government sources, has considerably improved for at least the short term. It is rare for any section of this fleet to average more than 250 days at sea per year, although there is proportionately less time lost in making the passage to and from the fishing grounds than with larger vessels.

Returns from Faroese distant-water trawlers and long-liners for 1964 and 1965 show that although the trawlers (which are bigger vessels) had twice the gross revenue of liners, the percentage return on capital for both was around 10 per cent. Liner earnings ranged from kr. 817,000 (£40,800) to kr. 1,904,000 (£95,200) with a mean of kr. 1,300,000 (£65,000), while those from trawlers ranged from kr. 1,271,000 (£63,500) to kr. 5,124,000 (£256,200) with a mean of kr. 2,730,000 (£136,500). Both sections of the fleet were used intensively, and the average number of days at sea was over 300 per year, although the time spent in making the passage two or three times a year to the main grounds in the north-west Atlantic cut down the fishing time of most vessels by between 30 and 50 days.

For the trawlers, pay to the crew was about one-third of the gross receipts, and for the line-boats rather more. For the latter bait and gear each accounted for about 10 per cent. of receipts and fuel about 5 per cent.; trawlers gear expenses were also about 10 per cent., but fuel here was over 12 per cent. Although essential for the Faroese fishing, ice and salt together were minor cost items at under 3 per cent.

The figures for crew remuneration in the Faroe Islands contain a considerable differential, and this is characteristic of capital intensive fisheries in Germany, Britain and elsewhere. The earnings of skippers were from three to five times those of deckhands on the averages for the two main sections of the fleet, and in the best trawlers were over six times. The responsibility on skippers is emphasized by the fact that they earned about twice the income of the next officers—i.e. first mates.

CATCH, PRICE AND PARTICIPATION IN FISHERIES:
The Examples of the Lofoten Cod and the Winter Herring in Norway

A relationship which is of basic importance in a fishery is that between the numbers of fishermen (or vessels) participating, and the yield of the fishery. For most countries, data on individual fisheries are not available in sufficient detail to show this, as such data tend to be compounded into gross annual returns. For Norway, however, the Lofoten cod and the winter herring fisheries have been of such dominant importance that they have been recorded in unusual detail for almost the whole of the twentieth century. There are yearly returns of catch, and of the numbers of fishermen participating; yearly average prices are also available, and although the Lofoten cod here are included with the other cod of northern Norway, it can safely be assumed that the all-over averages apply to Lofoten with substantial accuracy. Although for both these fisheries prices were brought under control by government marketing agencies in the inter-war depression and have remained so, the effect of this has been more to limit excessive fluctuations rather than to distort price levels. In addition, for the period since the Second World War characterized by inflation, there are separate price indices for cod and herring groups of fish. For both fisheries, the periods 1912 to 1926 and 1952 to 1966 are studied below. The first period features a situation where they were both labour-intensive, and in which the changes resulting from big price fluctuations (as a result of war-time conditions) were shown. This period shows the results of a vigorous stimulation of demand in war-time, followed by ensuing readjustments in more normal conditions. The second period illustrates the decline of traditional fisheries in the modern period, with the transition towards capital-intensive enterprises.

THE LOFOTEN COD FISHERY

In both these periods, the Lofoten fishery was basically traditional in its operation. A small size of craft has been general, and the average crew size has been consistently between three and four, while the equipment used consists of hand- and long-lines and gill-nets. The richness of the resource has itself here contributed to the survival of traditional methods. It may be noted that for both periods, of the three variables of catch, price and labour force, it is the last which is most stable and slowest to react to change—a common characteristic in fisheries. The size of the labour force in Norway has actually been abnormally flexible, as characteristically the men have been farmer-fishermen and have had partial support from their holdings which has been especially valuable during poor fishing seasons. It was established practice for men to come to Lofoten for the season from most of the Norwegian coast, and the volume of movement varied with the yield of the fishery. As a general rule, the effort varies with the size of the catch, and with the prices realized. The size of the catch, however, has little direct influence on the price, as Lofoten cod has been in competition with that from a much wider area of the North Atlantic, including especially Iceland and Newfoundland as well as the rest of Norway. The delivery of the cod in a preserved (dried, salted and now frozen) state to export markets fosters a degree of international competition unknown in fresh fish markets.

The period 1912–1926 (Figure 15) is useful in showing the reaction of the fishery to big price fluctuations, which were the result of the abnormal circumstances of the First World War. At this time most of Norway's fish were sold to Britain who, after some bargaining, agreed to pay high prices in exchange for a guarantee that she would receive 85 per cent. of the Norwegian export.[10] This led to some flow of labour into the fishery, although the 1916 war-time peak of fishermen was only 26 per cent above the 1913 figure, while the price rose four times. Poorer catches at the end of the war were associated with falling numbers of fishermen; thereafter, although prices in the 'twenties dropped from their war-time

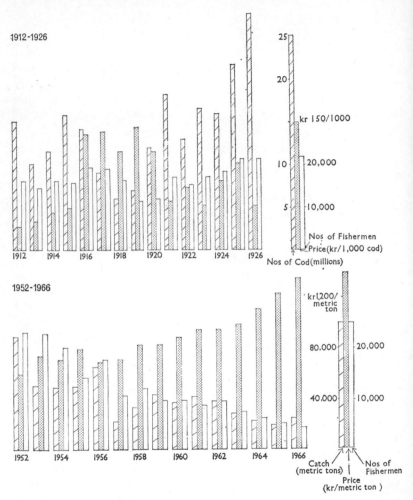

Figure 15. Relationship between catch, price and number of fishermen for 1912–1926 and 1952–1966 in the Lofoten cod fishery.

levels, they were still largely good, and big catches were associated with bigger numbers of fishermen.

There is some tendency for the level of participation in a year to be influenced by the circumstances of the previous one. The lower catch and price of 1918, for example, was followed by a reduction of 30 per cent. in fishermen in the next year, while the big catch and rising price of 1920 was followed by a 40 per cent. rise in fishermen in 1921. While the size of the catch varies in a largely random manner, there tend to be runs of good and of bad years, and an upward move in catch—especially when prices are favourable—encourages participation in the fishery.

Figure 15 also shows the changing modern situation of the Lofoten fishery. During the period (1952–1966) it has remained basically traditional in operation, partly due to political pressures, although the use of the more efficient purse-seine was allowed on a restricted scale between 1950 and 1958. A more powerful factor in accelerating the decline of participation in the fishery has been a reduction in the size of the fish stock, due to heavy exploitation of the cod during the part of the year which it spends in the Barents Sea and in migrating to and from Lofoten. Broadly, in this period there is an overall tendency for the catch per man to increase slightly, while the average price rose 128 per cent. between 1952 and 1966; in the same period, however, the price index for the cod group rose by 110 per cent, so that the rise in real value was minimal. In effect, the rise in productivity in the fishery has been small, and has certainly lagged much behind the general rate of economic development in Norway, with the result that better results accrue to labour in other sectors of the economy. The fall in participation, which amounted to 80 per cent. between 1952 and 1966, is mainly accounted for by a marked decrease in the volume of migration to Lofoten for the season; and for the fishermen who still come, there is a tendency to wait for reports of catches at the start of the season before they decide to engage. There is also considerably less recruitment now of younger men to the fishery in Lofoten itself and the age-structure in the Lofoten fishery revealed by the 1960 Norwegian Fisheries Census is markedly top-heavy. This decline in participation has been of some benefit to the men who continue fishing, as scarcer supplies have led to the average prices being above the guaranteed minima.[11] There is still some tendency for catch and price to influence the level of participation, as can be seen with the rise of fishermen in the good year of 1956, followed by a marked fall in 1957 when the catch fell disastrously, and there was some subsequent recovery in the better year 1958. This, however, is underlain by a deeper trend of general

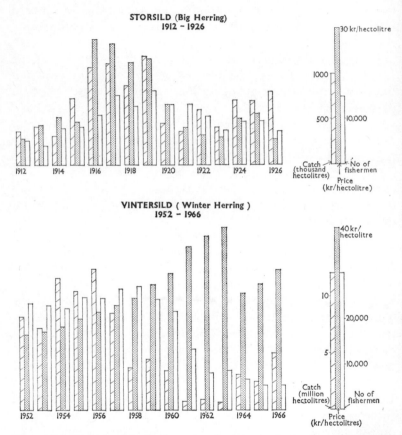

Figure 16. Relationship between catch, price and number of fishermen for 1912–1926 and 1952–1966 in the Norwegian winter herring fishery.

run-down, which is basically due to the increased impact on the fishery of other, stronger sectors of the economy; and this contrasts with the relatively stable conditions of the first period when—at a lower level of economic development—the fishery had much less competition for labour.

THE WINTER HERRING FISHERY

The Norwegian winter herring fishery shows greater fluctuations than does the Lofoten cod. A good deal of this is due to its being a pelagic rather than a demersal fishery, in which the catch is characteristically more irregular. Variations in effort are also greater, as a considerable proportion of the fishermen have had fishing as a secondary—rather than a main—occupation, which has been the case in Lofoten. In the period 1912 to 1926, there is a break-down of statistics between the two main parts of the fishery—the earlier 'storsild' (big herring) and the later 'vårsild' (spring herring). For the later period 1952 to 1956, the two figures are aggregated, but there are additional data on the capital value of the gear employed, and also on the number of units of different gear types.

Figure 16 shows the result of the stimulation of demand during the First World War in the 'big herring' fishery, and the subsequent readjustment to peace-time conditions. It can be seen that, although the war-time catch fluctuated considerably (within a range of 300 per cent. above the 1914 minimum), there was a flow of labour into the fishery, so that at the 1919 peak it was 350 per cent. above the 1913 minimum, with a marked fall in subsequent years. The year-to-year level of participation is also seen to be determined by a combination of catch and price; the fall in both, in 1920 for example, is associated with an 18 per cent. decrease in the labour force from 1919, while the recovery of catch and price in 1924 led to an increase in participation of 25 per cent. over 1923. The greater importance of the price than the size of the catch in influencing the level of participation in this fishery is instanced in 1913–14, and again in 1921–22 and 1925–26: in 1914 there was a fall in catch but a rise in price which was associated with a rise in fishermen, while in the latter two cases a rise in catch with falling prices was associated with reductions in the labour force.

The fishery was fairly stable in the inter-war years despite big price fluctuations in the 1930s. The period 1952-1966 shows initially a run of years of general stability, followed from 1958 by a period of disequilibrium. In this latter period

L

there has been a substantial reduction in the volume of herring coming to the Norwegian coast, while catching techniques have been revolutionized by the employment of bigger purse-seines, together with the power-block to haul the net. In addition, market outlets have been disorganized through the

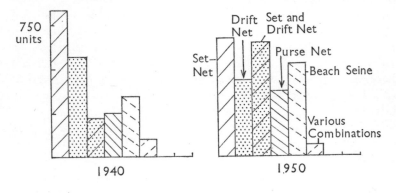

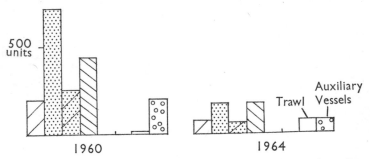

Figure 17. Trends in use of different gear types in the Norwegian winter herring fishery.

glut of the world fish meal market. Until 1957, there was a general increase in participation in response to a regular rise in price and an irregular rise in catch. At this period there was an expansion in the market; although the traditional salt-curing declined, there was a big expansion in reduction to meal and oil. The subsequent drastic slump in the catch to catastrophic levels in the period 1961 to 1963 was associated at first with a steepening of the price increase, but there was a big withdrawal of labour from the fishery as a bigger share of the catch was secured by power-block purse-seiners, and the number of fishermen decreased between 1959 and 1965 by 78 per cent. The greater detail of gear used which is available till 1964 shows clearly the move away from the simpler equipment (especially of set nets and beach seines) towards purse-seines and trawls (Figure 17); this has been accentuated since 1958 as very few shoals have come into the fiords where they can be taken by the simpler gears. The number of units was consistently well over 2,000 until that year, but by 1967 had decreased to 491. The more capital-intensive approach to this fishery, however, has scarcely succeeded in bringing it into adjustment with modern conditions. Although purse-seining with the power-block was initially very profitable, partly through extending the length of the operating year through engagement in other herring fisheries, it was rapidly overcapitalized by the construction of new craft and the conversion of old. This, along with big catches of anchovetta in Peru, led to the oversupply of the world fish meal market, with a resultant price fall; and subsequent poor seasons have led to the conversion of many of the craft to other fisheries or to outright sale. While the average price for herring rose by 140 per cent. between 1952 and the 1963 peak, the price index for the herring group rose 80 per cent. in the same interval. Since 1963 the real value of herring has been little above that of 1952, while the unit costs of operation have markedly increased. While the last decade has therefore seen capital investment, effort and yield in the fishery get radically out of adjustment, this has been mainly due to a natural fluctuation in the resource, though probably reinforced by over-fishing of herring in the North-East Atlantic. Readjustment has perforce been rapid through the withdrawal of capital and effort from the fishery.

In the post-war period, statistics of the value of the vessels and gear employed in the winter herring fishery are available until 1960. These, however, cannot be set against the value of the catch to indicate the return on capital, as the fishermen's income (i.e. labour costs) has come from the proceeds of the catch. The situation is further complicated in that the crews did much of the maintenance work on gear—and even boats—

themselves. It is none the less notable that in the years immediately after the Second World War, when the equipment consisted of old boats and gear, the value of the catch in good years might actually exceed that of the capital equipment used in its production. Subsequently investment in new boats and gear led to the catch value being substantially in the range of 40 per cent. to 50 per cent. of that of the capital equipment until the late 1950s. Since then the fishery has contracted and become much less labour-intensive; while no comparable statistics are available, it is known that the annual value of the catch is now a much smaller fraction of that of the capital equipment employed.

REFERENCES

1 Sir A. Hardy, *The Open Sea*, vol. II, 1959, pp. 224–25.
2 G. Borgstrom, 'The Atlantic Fisheries of the U.S.S.R.', in *Atlantic Ocean Fisheries*, ed. G. Borgstrom and A. J. Heighway, 1961, p. 303.
3 G. Borgstrom, *Ibid.*, p. 293.
4 *Economic and Social Development Program for Spain*, 1964–67, prepared by the Commissioner for Economic and Social Development Planning, Presidency of the Government of Spain, 1965, p. 124.
5 N. P. Sisoev, *Ekonomika Ribnoi Promyshlennosti*, 1966, p. 278.
6 G. Borgstrom, *Ibid.*, p. 283.
7 N. P. Sisoev, *Ibid.*, p. 279.
8 A. Holm, 'Norwegian Investigations into Costs and Earnings Pertaining to Fishing Operations', in *Some Aspects of Fisheries Economics*, I, ed. G. M. Gerhardsen, 1964, pp. 127–43.
9 J. R. Coull, 'Fishing in North-East Scotland', in *North-East Scotland: a Survey of its Development Potential*, H.M.S.O., ed. M. Gaskin, 1969, pp. 88–89.
10 O. Riste, *The Neutral Ally*, 1965, pp. 96–107.
11 M. Barbe, 'La Pêche aux Iles Lofoten', *Revue de Geographie de Lyon*, 1966, p. 53.

CHAPTER VIII

Ports and Landings

Ports and landing bases function as the points of transference of catches to markets and to land distribution systems. They are also very generally the points where stores to serve fishing fleets are gathered, and where necessary servicing facilities are concentrated. In effect, nearly all sectors of the industry —catching, selling, processing and distribution—are organized at the ports, and they are the essential nodes of the whole system. It is, however, rare for fisheries bases to have a completely specialized function, and the great majority also handle some general cargo; but fishing is the main activity of many ports, and the bigger ones generally have specialized fish docks.

Until modern times, there were considerable advantages in dispersal in fisheries operations, and it was usual for fishing to be on a small scale from large numbers of scattered points along the coast. One reason for this was the limited range of the majority of vessels employed, which were powered by oars or sail; another was that most fish which entered into commerce were preserved by curing (drying, salting or smoking), and there was little urgency in putting them on to fast transport to market. For centuries, however, there has been some concentration of operations at bases involved in long-range fisheries, such as the Dutch ports with their herring fleets, and the ports in various countries which engaged in the cod fisheries at Iceland and Newfoundland.

In modern times, the forces of concentration have become much more powerful, and have operated in all sectors of the industry. Larger catches taken by individual vessels and by fleets have demanded greater capacity at landing bases.

The increasing size and complexity of the vessels themselves have required deeper harbours and an expanding range of servicing facilities for unloading, fuel supplies, marine engineering and repair facilities, the supply of specialized fishing gear and electronic equipment. Increased capital investment has been required for processing, in canning, freezing, reduction and other uses; and there has been a concentration of transport facilities, by both rail and road, for the large part of the catches which go fresh to market. Once begun, the forces of concentration have tended to be self-sustaining, and an important factor has become the concentration of demand in the bigger centres; auction systems of selling have become general in most of Western Europe, and this has had the effect of raising general price levels and of attracting landings. Even where auction systems are general, smaller bases tend to offer lower prices to fishermen; this is related to fewer buyers and poorer transport facilities, but also to the frequent necessity of incurring the extra expense of dispatching the catches to main bases for processing.

In the relationships of fishing ports with their hinterlands, the main function of the ports is the distribution of the catches to consuming points. An essential ancillary function is the gathering, storing and selling of a series of materials essential to the fleets. This has always included such ships' stores as tar, cordage, and timber, and indeed the construction of fishing boats is usually carried out in the ports. It has included, too, supplies of gear like nets and lines, and stores of barrels and salt. Included now are fuel, marine engines, winches, and electronic equipment like echo-sounders and navigation aids. While there are always economic advantages in obtaining supplies near to bases, materials like salt and timber have for centuries been traded internationally. Now fishing ports have contacts with a great variety of centres to supply the range of stores required; and although the trade circulates mainly within national frontiers, a growing proportion of the more sophisticated and higher-value capital equipment now is traded internationally: such items include electronic equipment, gear items (like purse-seines), engines and vessels themselves. The contemporary movements towards economic integration in Europe are accelerating this trend.

FISH MARKETS & PRICE SYSTEMS

The main media through which fish enter the wholesale and retail sectors of trade around the coasts of Western Europe, from the west coast of Sweden round to Italy, are auction markets.[1] These have been established mainly to cater for the fresh fish trade; they have been found to be the best means of gearing supply and demand, and of concentrating landings at points for rapid dispatch inland. In the countries of Eastern Europe, and those of the far north-west, auctions are much less important. In the former group, there are still many scattered landing-points on the Baltic and Black Sea, and on the bigger rivers and lakes; but in the U.S.S.R., Poland and East Germany the large scale post-war expansion has been associated with planned centralisation of landings, processing and service facilities at main ports like Murmansk, Gdynia and Rostock.

In the Mediterranean zone, and to a lesser extent in France, the centralization movement has been less strong, or delayed in its effects. There are still many landing places, and associated with a large proportion of them are considerable numbers of small operators of limited mobility, although auction systems are the general rule. There has also been, however, the emergence of a number of major ports, mainly associated with longer-range capital-intensive fisheries. Vigo in Spain is now one of the biggest landing bases in Europe, while Boulogne, Lorient and Concarneau are large ports; and in Portugal the concentration of trawler operation at Lisbon and Oporto has made them outstanding bases. Fishing bases in Italy are particularly widely dispersed; some concentration is occurring at trawler ports like Leghorn and Chioggia, but these have only about 4 per cent. each of total national landings. The other extreme may be instanced in Denmark, where seven main bases have over 80 per cent. of all landings, and in England where the two main trawler ports of Hull and Grimsby have over 75 per cent. of all demersal landings.

In Norway, Iceland and the Faroe Islands in the north-west, centralization of operation and landings have been much less marked than in most countries, despite the scale of their effort in fishing and the great importance of fisheries in their economies. This is linked to the systems of fixed prices that have been adopted, in situations where the bulk of catches are exported, and where the influence suppliers can exert on the market is limited. There is usually a longer delay between landing and consumption, and higher processing costs are often incurred. The price systems involve the fixing of prices in advance for individual seasons, in relation to forecasts of production and of market conditions. They do give more stable conditions for the fishermen, but they lead to complicated regulations to give flexibility to cope with seasonal fluctuations, costs of gathering from different landing points, and ultimate dispatch to a variety of destinations. Because of the extra costs of gathering catches and of export, the prices to catchers, although guaranteed, are considerably lower than the average in countries with auction systems, and must be made good by greater turn-over, although this is frequently possible from the rich Arctic grounds. In Norway, fishermen's co-operatives largely control price levels, and in Iceland and the Faroe Islands prices are fixed by boards which have representatives of fishermen, vessel owners and fish buyers. In these countries, the main bases have in effect developed as a special type of entrepot: direct landings in them are often relatively small, but fish are collected from a variety of smaller landing places, and occasionally from distant-water vessels. Examples of them include Ålesund, Kristiansund and Svolvaer in Norway, Reykjavik and Seydhisfjördhur in Iceland and Klaksvik in the Faroe Islands. Processing and storage are often concentrated in such centres, and they are also the main locations of marketing organisations.

While there has been some concentration of operation from particular points in Eastern Europe, this has occurred within a different framework of organization. Fish prices for all ports are determined at national level, and this has led to the increase in landings at dispersed points on (for example) the Baltic coast of the U.S.S.R., by vessels belonging to fishermen's collectives. At the same time, the main part o í the expansion has been achieved in the state sectors of the fisheries, in which planned investment has been concentrated in selected centres like Murmansk, Kaliningrad and Gydnia.

Auction systems have proved to be suitable for the rapid

disposal of catches of mixed fish, and there is automatically some quality grading in that merchants pay higher prices for fish in better condition. The gearing of supply and demand is, however, imperfect and in several countries measures are taken to protect producers by preventing prices from falling to uneconomic low levels, which are particularly prone to occur in summer with slackness in demand. These measures in effect set minimum prices, although often (as in the Netherlands and France)[2] they are financed from within the industry itself. In Belgium, the government makes good any shortfall; and in Spain a special body is empowered to intervene when prices are in danger of falling below a set minimum, and also attends to distribution.[3] Minimum prices administration can be complex and difficult, however, and Denmark and Italy operate without them; and in Denmark export outlets are sufficiently stable to maintain price levels fairly reliably. In Britain minimum price arrangements have largely been limited to schemes of merchants at a few ports, although a national scheme for white fish was agreed in 1970.

Single-species fisheries, like herring and sardine, present their own problems in selling. This is partly because supplies are more irregular, and the amplitude of price variation greater: within days they may fluctuate through several hundred per cent. More government regulation is frequent in such fisheries, and in Britain it has been usual for the Herring Industry Board to set particular minimum prices on herring used for different purposes. Another problem with pelagic fish is that landings are often in greater bulk and the display of the whole catch on a market floor more difficult, and in several countries sales are based on samples of catches.

The institution of auction markets which best suited the organization of the industry and the transport systems in the hey-day of the railway is not now so suitable for the trade, although it still has special advantages for the fresh fish sector. However, with the integration of enterprises—both horizontally and vertically—which is characteristic now, the auction hinders rapid transfer at a vital stage for vertically integrated firms. There is therefore some tendency to by-pass the auction —especially for fish frozen at sea, where it is particularly important that an unbroken cold chain be maintained right through to the consumer. In such cases contract prices may be fixed in advance, and the uncertainties of the auction eliminated.

CENTRALIZATION OF FISH LANDINGS:
The Example of Great Britain

The centralization of port functions associated with the fishing industry during the modern period may be illustrated from Britain, where the rise of the main trawling ports has dominated development[4]. This trend developed earliest in Britain and Germany, and has proceeded furthest in them. It can be best traced in Britain because of the existence of unbroken statistics over a longer period, and also because of the complication of the post-war division of Germany. It is for demersal fish, which have been the dominant sector of the British market, that the effects of centralisation of operation have been most marked. This has been partly due to demersal fish being less perishable than the other main types (herring and shell fish), and hence able to be carried longer distances to landing points by the relatively slow medium of sea transport. Centralization has also affected demersal fish rather than other types (particularly herring) because they are available all the year, and have been better able to justify more overhead expenditure on port facilities than those varieties of which the landings on particular parts of the coast are limited to restricted seasons. Even so, much of the national commerce in other types of fish than demersal has come to be organized from the main trawling ports, and a bigger share of processing is concentrated in them than is suggested by the size of their landings.

Available government statistics can illustrate the centralization trend over nearly all the relevant period in Scotland, where figures are available from 1887. In England, however, the trend gathered momentum earlier, but government statistics do not begin till 1903. Both, however, show the effects of the freer and more rapid access to big (and growing) markets that came with the railway. They also show the later effects of the growth of the more flexible medium of road

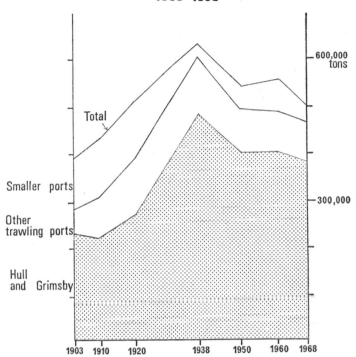

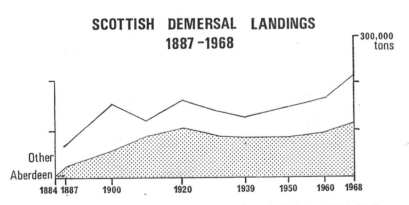

Figure 18. Centralization of demersal landings in Britain, 1884–1968.

transport, which has added to the ramifications of the marketing system.

The 1903 situation in England (Figure 18) shows that by that date the concentration on the main ports was substantially complete: already over 70 per cent. of all demersal landings were at the six main trawling centres (Grimsby, Hull, North Shields, Lowestoft, Milford Haven and Fleetwood), and 58 per cent. of the English total was on the two Humber ports alone. While comprehensive data are not available for earlier periods, it is clear that concentration at particular ports had been going on in the century prior to the first collection of official statistics, and that at first the main developments were in south-east England from such ports as Ramsgate and Barking. Here fishing was carried out from line-smacks and trawlers, and improved roads gave access to the big metropolitan market. A second and much more important stage in concentration can be distinguished from the mid-nineteenth century with the coming of the railway, which most stimulated the two Humber ports in the first instance. Grimsby, which has been the premier port virtually throughout the modern fishing history of Britain, was in particular largely a creation of the Manchester, Sheffield and Lincolnshire Railway Company. As the scale of operations rapidly expanded, the Humber ports were better favoured than the earlier ones in the south-east, as their location in relation to the configuration of the North Sea makes a much greater area of fishing grounds within practicable distance for exploitation. The rise in landings came first from sailing trawlers and line-smacks, supported by fast transports to help get the fish fresh to market, and by the 1870s some of the transports were steam-powered. The number of vessels operating from Grimsby rose from 40 in 1860 to 587 in 1880; and the landings from 3,435 tons in 1857 to 36,784 in 1875. Such striking figures indicate rapid investment and rise in productivity in boom conditions, with a transformation of the national pattern of activity in fishing, as well as a great upsurge in the national market for wet fish. The expansion of the port continued vigorously to the end of the century: there were 1,118 vessels worked from it in the peak year of 1893, and by the early years of the twentieth century the annual catch exceeded 150,000 tons. These spectacular developments were accompanied by, and indeed necessitated, the establishment of a series of specialized facilities at Grimsby— the building of fish docks and the fish market, coal and ice stores together with other marine supplies, and marine engineering establishments. Similar facilities were to follow at other main ports.

On the national plane, a second phase of centralization came from the 1880s; from this time the more powerful and more mobile steam trawler became the main catching vessel, and the other five main trawling ports were established (North Shields, Lowestoft, Milford Haven, Fleetwood and Aberdeen). These provided landing points scattered around the coasts of most of the more populous parts of the country, and were able to deliver fresh fish supplies by rail to a market which by 1900 had national dimensions. Landings were also increased at several other ports which did not survive the competition in the longer term; these included mainly Boston, Cardiff and Liverpool. Auctions became usual at these main ports, replacing earlier systems of fixed and contract prices from merchants, and the higher general levels of prices obtained were themselves incentives to continued expansion.

The late nineteenth century development of trawling occurred in the main with the different ports exploiting the particular sea areas adjacent to themselves, and acting as the transfer points for the resources of these areas to reach landward markets. The rapid transport to market provided by the railway was a factor leading to considerable overlapping in the landward spheres of the various ports, and indeed all the trawling ports (including the most remote one, Aberdeen) developed regular consignments to the metropolitan London market. Patterns of branching distribution from each port evolved, and important points in the system were secondary distribution centres of inland wholesale markets in major cities like Manchester, Birmingham and Leeds, as well as Billingsgate in London.

The early twentieth century saw further important developments in catching with the growth of distant-water trawling in Arctic waters, particularly from the leading centres of Hull and Grimsby. This was spurred on by the bigger catches from these areas, together with decreasing yields from the nearer grounds as a result of the beginning of over-fishing; and it was made possible by the trawler having developed to a size and efficiency which made long-range operation feasible. The increasing concentration on Arctic grounds in the inter-war years allowed total national landings to continue to rise, while an even greater proportion of them became concentrated at the trawling ports. By this time the profitability of near- and middle-water trawling was somewhat adversely affected: here there were lower yields while costs increased, and the national fish market by this time was practically saturated, while general income levels had risen to standards which led to proportionately more being spent on other sources of animal protein. This led to the thinning out of the English trawl ports to the six main ones above, and in addition direct landings at Billingsgate were suspended after 1937, largely because of the congestion of shipping in the Thames estuary. The overall result was that by 1938, 96 per cent. of all English trawl landings were made at the six main ports; and 77 per cent. of the total was at the two Humber ports, which were the only ones to increase their landings over the inter-war period.

The latest phases have witnessed a limited degree of decentralization in the pattern of landings with the advent of the more flexible medium of road transport. This began in the fish trade in the inter-war years, but its main impact has been felt since 1950, and it has been accompanied by increased landings from improved inshore fleets (employing light trawls and ground seines) at ports such as New Lynn and Whitehaven. The greater emphasis on quality in the modern market has been an advantage to fleets which can make daily landings, while a combination of increased costs and more intense international competition on grounds exploited by the distant-water sector has somewhat decreased its importance. The rise of road transport has also stimulated a more complex system of distribution from the ports, as an increasing pro-portion of the fish have been sent directly from the ports to inland retailers, by-passing the inland wholesale markets.

In Scotland, the pattern of concentration produced within the last century is simpler and can be more completely demonstrated. Although trawler fleets were operating from several points on the Scottish coast at the end of the nineteenth century, only at Aberdeen and Granton did they make a lasting impact, and the former has been much the more important. Steam trawling began from Aberdeen in 1882; it was supplemented to a limited extent by great-lining from steam vessels, and there was a period of vigorous expansion until the First World War, during which investment paid high dividends. In 1884 (the first year of available records) more than 4,500 tons were landed, and by 1887 this had risen to 22,310 tons, which represented 32 per cent. of all Scottish demersal landings. In the years before the First World War, landings at the port were well over 75,000 tons, or over 70 per cent. of Scottish landings.

Subsequent developments in Scottish fisheries have decreased the relative—but not the absolute—importance of Aberdeen. The main reason for this was the rise of ground scining for white fish, first at ports in the north-east, and subsequently extending round the whole coast. This was stimulated in the first instance by the large inter-war contraction in markets for Scottish herring, which resulted in a growing proportion of the big herring fleet turning over to scining. Although a considerable proportion of the catch of the sciners is landed at Aberdeen, the post-war period especially has seen the rise of landings at a series of widely distributed points from which the catches reach market by road. By the 1960s, the proportion landed at Aberdeen had fallen to barely 50 per cent. Although about two-thirds of the Aberdeen landings are dispatched to consumption points in England, it is mainly road transport which is now employed for the purpose; only for the longer haul to London and south-east England does the more rapid rail transport operate.

If the share of direct landings at the main trawling ports has tended to fall in post-war years, it still accounts for by far the major part of the catch. They also control a bigger share of the total than is apparent, as overland consignments from smaller ports for first sales augment direct landings at the main ones, where prices are generally higher in more competitive markets. Agents of firms based in the main ports buy considerable proportions of the catches landed elsewhere, and these are often sent to the bigger centres for processing. This is especially notable in Scotland, where the main market and processing centre of Aberdeen collects supplies from a wide area in the north and west of the country.

TRENDS IN FISH PRICES

The economic viability of fisheries enterprises has been seriously affected in the post-war period by inflation. The modern tendency is for the rate of rise in price of primary products to be slower than that of manufactured goods, and this means that prices for fish have generally lagged behind those for the capital equipment required to catch them. Although there has been some compensating increase in turn-over in fisheries enterprises with larger catches, this has often not been sufficient to cover rising costs.

Prices for fish for human consumption have generally increased at least double over the post-war period, and are now often around five times the levels of 1938. In countries in which currencies have been less stable the rise has generally been greater, and the most prominent example of this is that of Spain, where the average unit value in 1968 was almost twenty times that of 1938. Iceland, too, has suffered from excessive inflation, although no precise values for total catch over the period are available. In the countries that have developed big fisheries for reduction in the post-war period, the low values realized have limited average increases in unit value, and in Denmark the unit value actually decreased 37 per cent. between 1948 and 1968. Here the considerable increase in catch even for food fish has tended to produce the conditions of a buyer's market, and prices of most of these fish in 1968 were two-and-a-half to four times those of levels in the 1930s. On the other hand, values of cod landed in Norway increased eight times over the same period, partly because of the increased proportion used for freezing.

Although there has been a fairly consistent upward trend in most fish prices in the last thirty years, the prices of boats and equipment over the same period have frequently risen ten to twenty times. While prices of the most prized fish varieties— salmon, lobster, nephrops and some other crustaceans and molluscs—have risen over ten times, these represent only a minor proportion of catches. Labour costs, too, have risen more steeply than fish prices, and have contributed to the somewhat precarious economic position of the fishing industry in many countries.

AREAL VARIATIONS IN FISH PRICES

Areal variations in fish prices within Europe are considerable, and depend mainly on the location of landing places in relation to major consumption centres, and to the costs of processing and transport of a perishable product. Various regulations on the fish trade—mostly at international level—have also conditioned price levels at particular landing points, although the present trend is for more free trade in fish. The depressed level of prices in the main exporters of the north has already been mentioned; another international phenomenon is that fish prices tend to be relatively high in countries in which other food prices are high, such as France and Italy.

The factors that cause fish prices to vary spatially can be more clearly seen in the simpler conditions existing within national markets. In Britain, for example, the smaller distance to major consuming centres from the main trawling ports of Hull and Grimsby, allied to greater concentration of demand from port wholesalers, raises price levels above those of the more remote northern port of Aberdeen for comparable qualities. The same effects are enhanced when landings are made at island ports, and catches are subjected to extra trans-shipment costs to mainland destinations: in Denmark, for example, the prices for cod on the island of Bornholm are c. 20 per cent. below the national average, and those for plaice c. 50 per cent. This is even more prominent in Portugal with the landings of some species on the distant islands of Madeira and the Azores; landed values of tuna on the Portuguese mainland, for example, are between three and four times those on the Azores.

VARIATIONS IN FISH PRICES IN SCOTLAND

Areal variations in fish prices may be illustrated in greater detail by the average prices for demersal fish and for herring

M

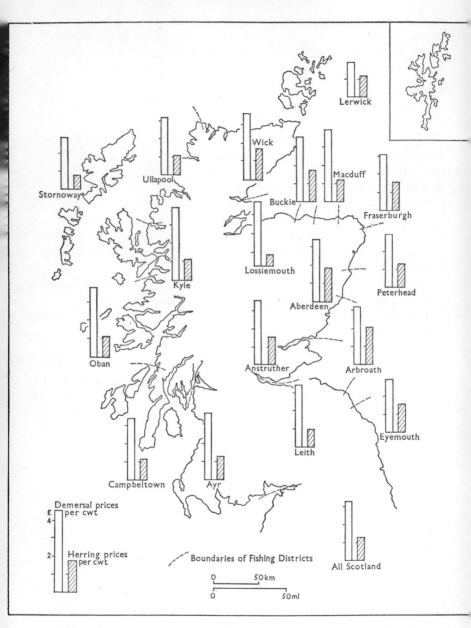

Figure 19. Average fish prices for the 1958–1967 period in the fishery districts of Scotland.

in the different fisheries districts of Scotland, where they dominate the total catch. The general relationship of price to distance from consuming centres is fairly clear, but there are two complicating factors. One is that the composition and quality of catch varies in some measure between districts; the other is the practice of consigning part of the demersal catch from some districts to more competitive markets for first sales as in these circumstances the value ultimately realized is recorded at the place of landing.

Figure 19, which shows the average prices for the two kinds of fish for the ten-year period 1958–67, is basically a case of a 'core and periphery' relationship. The core embraces the most populous part of Scotland in the midland belt between Glasgow and Edinburgh, but effectively stretches north-eastwards to include the main fishing port of Aberdeen. The periphery includes the coasts of the Highland area of Scotland, and extends out to the Northern and Western Isles. The market for fish in Scotland is not, however, a closed one: a considerable proportion of both white fish and herring (approximately one-third) is sent to England, and this includes the consigning for first sales of the great part of the landings from the Leith district to the major national fishing ports of Hull and Grimsby. Part of the herring also goes to continental destinations, and (in winter) some herring comes from Norway.

Demand for landings is greatest at the main port of Aberdeen, and this is the destination of most consignments of demersal fish made in Scotland, especially from the north and north-west. Consignments are also made to Glasgow, where the concentration of demand at an inland wholesale market is sufficient to attract them.

Prices of Demersal Fish in Scotland

Aberdeen is by far the biggest market for demersal landings, taking approximately one-half of the total catch. The concentration of demand associated with this raises prices, but the average price (at £3.42 per cwt.) in Aberdeen is not the highest of the fishing districts, because of both quality of landings and transport costs to market. The port is the main centre of trawling, and the major part of the landings comes from middle-water trawlers which spend 10 to 15 days at sea per trip. Consequently, although of better quality than the distant-water fish which dominate the English market, they are less fresh than catches from inshore fleets. In addition, some two-thirds of the Aberdeen landings are subsequently subjected to the costs of the relatively long haul to consuming centres in England. The prices on the coasts around the Scottish midland belt are generally at least as high as those of Aberdeen, despite the lesser competition in demand in coastal markets. Prices in the Leith district, which includes Granton, the other trawling port of Scotland, are on a par with those of Aberdeen. Although the quality is similar, and the competition of demand is less, Granton is sufficiently close to the Humber ports for the bulk of the landings to be consigned there for first sales; and this compensates for the rather lower prices realized at Granton itself. While lesser competition of demand depresses the average price levels in the Eyemouth and Arbroath districts in this region, the landings are prime quality inshore fish, and this, linked to more competitive buying is responsible for the prices in the Anstruther and Ayr districts being above those of the trawling ports. This is more noticeable in the Ayr district in which landings are concentrated in its northern part close to the big consuming centre of Glasgow.

The influence of transport costs from the periphery can be seen in the lower average prices obtained (£3.21 per cwt.) in the Ullapool district in the north-west. It is better shown, however, in the islands, from which marketing costs are inflated both by steamer freights and by trans-shipment costs. In the Lerwick (Orkney and Shetland) district, prices at £1.93 per cwt. are 44 per cent. below those of their mainland terminal at Aberdeen. The less pronounced depression of prices in the Stornoway (Outer Hebrides) district is related to much lower catches being landed substantially for a local market, and not being subjected to steamer freights.

The remaining districts are in the north and west of Scotland, and they illustrate again the effects of the composition and quality of the catch, together with some results of overland consigning. The relatively low figures in Peterhead and Fraserburgh basically reflect landings dominated by bigger seine-net boats which aim for bulk rather than quality, and which generally spend a week at sea at a time. Landings in the other districts, although generally of smaller volume, are of prime quality, and are mainly made after one-day trips. Around the Moray Firth and in the Kyle and Oban districts of the west coast, too, landings have a higher proportion of the more valuable flat fish (such as plaice and sole) among the cod, haddock and whiting that dominate Scottish demersal landings. Average prices (at over £4 per cwt.) are highest in the whole of Scotland in the Macduff and Kyle districts. The practice of consigning catches to Glasgow, especially from the nearer Campbeltown and Oban districts, also helps to raise the averages for west coast districts.

Prices of Herring in Scotland

Despite herring being in principle a single species, average values for them show a wider proportional range than for demersal fish. The main reason for this is the relatively high level of transport costs which is due to the average value per cwt. for herring being less than 40 per cent. of that for demersal fish. This in turn is related to herring being a shoaling pelagic fish which tends to be caught in considerably bigger numbers per unit of effort, and also to a sector of the herring market being for low-value uses including curing and reduction to meal and oil. Another reason for the variation in herring prices is the quality of the herring themselves. Winter and spring herring are often lean and of poorer quality; they may be of a smaller size, and may be mixed with the lower value sprats.

The location of processing plants is of considerable consequence to the levels of herring prices. The main concentration of processing plants is in the north-east, especially in Aberdeen but also in the Peterhead and Fraserburgh districts. Herring are brought here from all over Scotland for processing, and prices (from £1.29 to £1.82 per cwt.) in this area are notably higher than those of the west coast (below £1.20 per cwt.), from which much of the catch is transported by road to the north-east for processing. For herring landings in the Lerwick district (Shetland) a high proportion is processed in the islands, and the islands with an average price at £1.18 per cwt. are not in this case at the same disadvantage as they are in demersal fish prices, even if much of the processing is for the lower-value uses of pickle-curing and reduction to meal and oil. In this case, the lowest prices are realized in the Hebrides (Stornoway district) where the additional cost of steamer transport depresses the level to £0.76 per cwt.

The volume of landings is itself a big factor in the determination of price levels; in some districts herring landings are small and sporadic, and districts like Eyemouth and Arbroath have only around 1 per cent. of the total landings of major districts like Fraserburgh and Ullapool. Where landings are low a big proportion often goes to the best sector of the market—i.e. are sold fresh; and despite the lack of competition at first sales for such landings, average prices are among the highest. This is the explanation for the high values at Arbroath (£2.04 per cwt.), Anstruther (£1.56 per cwt.) and Eyemouth (£1.56 per cwt.), which are favourably placed for transport in the most populous central belt, and it contributes also to those of Wick and Buckie.

The cause of the biggest price variations is, however, variations in quality. This contributes to the lower values generally realized on the west coast, where the main part of the landings are in winter, and the herring generally leaner and lower in fat content. Even in the Ayr district, which is favourably placed for the big Glasgow market, the prices (£1.25 per cwt.) are relatively low. The lowest values of all, however, are found in the Lossiemouth and Leith districts. In both of these the fisheries are in winter and spring, the herring are small and often mixed with sprats, and a particularly high proportion of the catch goes for reduction to meal and oil.

THE RELATIVE IMPORTANCE OF DIFFERENT PORTS

Fishing ports vary in the scale and importance of their activities, and this can be seen at continental, national and regional levels. Initially most fishing bases depended on the existence of local grounds, but the major ports have emerged as the result of the exploitation of grounds at greater ranges;

and the extent to which they have developed has frequently depended on the initiative of various individuals and organizations, rather than decisive natural advantages. The distribution of main bases could scarcely be claimed to conform closely to any economic models or location theories. For the most part they have developed as responses to needs within individual nations, and only recently have moves towards economic integration at the supra-national level begun to make any changes.

In the full assessment of the relative importance of fishing ports, it would be necessary to compute the total turn-over of all fisheries enterprises in each port. Such data are never in fact published, and even estimates for any port would be difficult to arrive at. The most direct indicators of the importance of different ports are volumes and values of landings; the ports with greatest figures for these in fact tend to have even greater status because of the concentration of processing and of services in them. Another useful indicator may be the numbers and tonnages of fishing vessels registered at each port: thus in West Germany, the leading port of Bremerhaven has 55 per cent. of the national total of deep-water trawlers, and 60 per cent. of their tonnage, and the deep-water fleet now very much dominates West German fisheries. Bremerhaven is also the most important base for luggers, although the smaller cutters are widely distributed along the coast. In Britain, distant-water trawlers, the main sector of the fleet, are almost entirely concentrated at the two main ports of Hull and Grimsby, while middle-water trawlers operate also from the six other ports which are next in importance in Britain. Numbers of vessels registered at a port may, however, be an uncertain index of its importance. Major ports tend to attract landings from boats registered elsewhere, and in seasonal fisheries boats may operate for several months of the year away from their home base. Thus Skagen and Hirtshals in Denmark have big landings for reduction from Swedish vessels; Aberdeen attracts many vessels registered at various ports on the east coast of Scotland, and Göteborg from boats from various points on the west coast of Sweden.

In most countries, it is possible to distinguish at least three or four levels of importance, from small landing places for a few inshore vessels, to the major bases with distant-water fleets and a complete range of ancillary services. In the main exporters of the north-west (Norway, Iceland and the Faroe Islands), however, there has been a less prominent emergence of main bases, despite the great importance of fisheries in their economies. The auctions that have promoted centralization in many countries are absent, and in Norway and Iceland much of the fisheries effort is dispersed along extensive coasts, part of it still in relatively small boats.

Total landings by ports are regularly published for only a few countries, but the major ports (by landed tonnage) at the continental level appear to be Murmansk and Kaliningrad in the U.S.S.R., Esbjerg, Skagen and Hirtshals in Denmark, Bremerhaven in West Germany, Hull and Grimsby in Britain, Ijmuiden in the Netherlands, Boulogne in France, Vigo in Spain and La Luz in the Spanish possession of the Canary Islands. In 1960, the Murmansk processing combine produced 300,000 tons of fish, and this must be equivalent to a greater landed tonnage because of weight losses in processing. This was scheduled to reach 350,000 tons by 1965,[5] and trends in national landings suggest that this was achieved. The only port in Europe which has exceeded Murmansk in landed tonnage is Esbjerg, where the total exceeded 500,000 metric tons in 1968, but the great bulk of this was for reduction and was of much lower value. Hull, Grimsby, Bremerhaven, Skagen and Hirtshals have landings in the 200,000 ton region, although the values at the Danish ports are again much lower. In France, Boulogne has annual landings of around 130,000 tons. In Spain in 1961, La Luz (i.e. the fishing harbour of Las Palmas) in the Canaries had landings of 114,000 tons, and Vigo of 74,000 tons;[6] but developments have been rapid and by 1965 the tonnage at Vigo had risen to 140,000 tons[7] as it has become the main base for distant-water operation.

At a lower level, there is a considerably bigger number of ports of great importance with landings in the range of 50,000 to 100,000 tons. These include, for example, La Coruna, Passajes and Algeciras in Spain, Aveiro and Lisbon in Portugal, Lorient and Concarneau in France, Tyborön

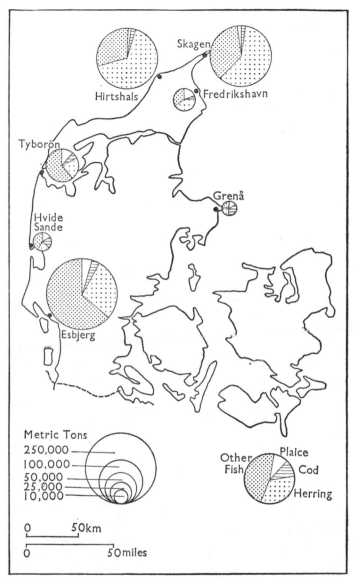

Figure 20. Average landings at the main Danish ports by species, 1963–1967.

in Denmark, Cuxhaven in Germany, and Aberdeen in Scotland. It is likely too that the main Polish ports—Swinoujscie and Gdynia—are of at least this rank of importance. The numerous ports at which landings are less than 50,000 tons are in fact frequently more dominated by fishing in their economies than the bigger bases, at which landings made in fish docks are only one of several main activities. Emden in West Germany, Ondarroa in Spain, Brixham and Peterhead in Britain, Ångholmen in Sweden, and Matozinhos in Portugal are examples where the port functions are dominated by fishing.

As an example of a country with ports which show a complete range of importance, the case of Denmark may be examined. Figure 20 shows the average landings for the period 1963–67 at the main ports in Denmark. Here landings have become progressively centralized, especially in the post-war period, and over 80 per cent. of national landings are made at the seven main ports shown, while over 65 per cent. of the total are concentrated in the three ports of Esbjerg, Hirtshals and Skagen. As well as differences in the overall importance of the ports, Figure 20 shows differences in the proportions of species landed. In all, Esbjerg is the outstanding port, and indeed has been so throughout the modern period; over the five year period average landings were 300,000 metric tons, and average landings at both Skagen and Hirtshals were around 220,000 metric tons. The port of Tyborön represents a third level of importance with average landings of nearly 70,000 metric tons, while the remaining three—Hvide Sande, Frederikshavn and Grenå have figures of 20,000 metric tons or less.

The different proportions of species in the landings, which influences the value at each port, is related partly to location. The dominance of herring at Skagen and Hirtshals is related to their proximity to the Skagerrak grounds, and the use of these (along with other species) for reduction, results in the value of catches being only about one half of those of the main port of Esbjerg. Plaice is prominent in the landings at the west coast ports, and the unit value of catches in Hvide Sande is highest of all Danish ports. The relative importance of cod is greater at Grenå on the east coast of Jutland, although landed tonnage is lower here than at the neighbouring port of Frederiks-havn, where landings are dominated by herring for reduction.

REFERENCES

1 O.E.C.D., *Price Systems in the Fishing Industries*, 1966, p. 12.

2 European Economic Community, *Rapport sur la Situation du Secteur de la Pêche*, 1966, Part II, pp. 172, 176.

3 O.E.E.C., *Fish Marketing in O.E.E.C. Countries*, 1952, p. 14.

4 See in this context: G. L. Alward, *The Sea Fisheries of Great Britain and Ireland*, 1932, pp. 130–370.

5 R. A. Helin, 'Soviet Fishing in the Barents Sea and North Atlantic', *Geog. Rev.*, 54, 1964, p. 402.

6 *Economic and Social Development Program for Spain*, 1964–67, 1965, p. 125.

7 P. Brady, 'The Emergence of Vigo as a Premier Fishing Port', *Fishing News International* 5, 1966.

CHAPTER IX

Processing

The importance of processing in the fishing industry is three-fold. In the first place, it generally serves to preserve a very perishable product; this allows it to be used for a longer time after catching, and to be transported over greater distances, and so widens the market. Secondly, processing reduces the weight of the catch, and reduces effective transport costs; and in the third place, there is a direct economic importance in processing in the value that is added to the product. A substantial part of the European catch is marketed fresh and unprocessed, and the necessity for processing depends largely on the time interval between catching and consumption. Processing is essential for the part of the catch not used for food, and the reduction sector in Europe is now second only to the fresh market in the proportion of the total landings it absorbs, although only a few countries in the north-west operate on a big scale in reduction.

Processing can be done either ashore or afloat, and distance of the grounds being fished from base is a major factor in determining which is employed. Although processing at sea makes processing units more mobile, and can increase the fishing time available to catching units, it is generally much more costly to build and operate factory ships than factories on land. Consequently, processing at sea is largely restricted to distant-water operation, although some of the simpler processes may be employed at lesser ranges. It is common for demersal fish to be gutted at sea even in inshore operation, and at middle- as well as long-range operation, herring may be gutted and pickled. At these ranges, too, crustaceans (such as shrimps) may be boiled and iced, and fish livers boiled to extract their oil. Even with investment in the more

elaborate machinery for processing in distant waters, there
are problems, as the installation of filleting and freezing plant
on distant-water trawlers has been known to cut the pay-load
by more than 50 per cent. because of the reduction in hold
space.[1] When factory ships are supplied by catchers, a major
problem becomes trans-shipment at sea, especially on Arctic
grounds with their frequency of gales. Processing at sea involves
principally salting and freezing, with canning of subsidiary
importance. Reduction is operated only as a secondary
by-product industry, as it would not be economic to incur
the additional expense of producing low-value commodities
afloat.

The precise proportions of the total European catch that
are allocated to different uses and processes cannot be stated
with precision, as comprehensive records are published for
only a dozen countries, and data are partial or absent for
others. If certain details are lacking, however, the major
elements of the picture can be discerned.

The fresh market sector dominates in the parts of Europe
where most fish are caught in local waters and where the range
of distance to inland consuming points is limited. It also
plays a big part in the heavily populated industrial countries,
even for catches from distant waters. In inland countries
like Hungary, the total catch is usually marketed fresh;
and in the Mediterranean zone the fresh sector of the market
dominates and accounts for 75 per cent. of the catch in
Greece and Italy. Finland and Ireland also draw the bulk
of their supplies from local waters, and in both 67 per cent.
of their catches are consumed in the fresh state. For several
of the nations that have fleets operating in middle and distant
waters, most of the catches are brought back unprocessed
on ice; and because of this the fresh sector still dominates the
market in Belgium, France, the Netherlands, Spain and
Britain. The fresh landings in Spain have consistently
expanded, and are now the highest in Europe at over 850,000
metric tons per year; those of Britain were formerly over
900,000 metric tons, but are less than 650,000 metric tons,
as an increasing part of the catch has been used for freezing,
as is the trend now in most countries. For the U.S.S.R.,
Poland, East and West Germany, which also operate

extensively in distant waters, smaller fractions of the catch are marketed fresh, and the major part is preserved by freezing and salting.

Even for fresh landings, the trend from the inter-war years was for filleting to increase: this decreases the weights and bulk for overland transport and gives a product which is more attractive on the market. This has proceeded furthest in West Germany, where 80 per cent. of all fresh fish are now sold as fillets[2]: most of the filleting is done at the main ports, but an increasing proportion is done at sea.

Fisheries for reduction now dominate production in several of the major fishing nations of Western Europe, but in Eastern Europe the modern expansion has been very much directed to increasing food supplies, and reduction is only a by-product industry. In both Norway and Denmark, over 70 per cent. of the catch is now used for reduction, and the reduction sector also tends now to dominate Icelandic fisheries, although the proportion of landings employed has varied sharply with fluctuations in the annual catch. Most of the Swedish catch also now goes for reduction, as in addition to the 25 per cent. of landings in Sweden so employed, most of the 49 per cent. recorded under 'miscellaneous uses' consists of landings made in Danish ports for reduction. In most other countries fish for reduction mainly represents an industry to utilize fish offal, fish of unsatisfactory quality, and surplus landings: it employs only 20 per cent of the West German catch, 10 per cent. of that of Spain, and 7 per cent. of British landings. As fish for reduction have the lowest value in the market, this sector tends to be supplied only after all others. The consequence has been that it has been particularly difficult to gear processing capacity to supply. This problem has been noted, for example in the Norwegian industry,[3] although it has tended to become less acute with the modern expenditure of much fishing effort specifically for reduction. Fish meal plants were begun as early as 1917 in Iceland, but the main expansion has been post-war, and there were 40 plants in 1959.[4] In Norway, reduction plants were also first established early this century, and the post-war period has seen a slight reduction in their numbers but a big increase in aggregate capacity; there were 64 plants in 1962.

Curing of fish is the main traditional method of preservation, and includes drying, salting and smoking. Over most of the continent these techniques are of declining importance, and demand tends to concentrate on products preserved by more modern techniques. Most curing is now done by the Eastern European nations with their distant-water fleets, and some of their factory ships concentrate completely on salting, especially in herring fisheries. In these countries the total tonnage cured rose until the early 1960s, since when it has somewhat declined and freezing has rapidly expanded. In 1968, 737,000 metric tons were cured in the U.S.S.R., and this was a reduction of 25 per cent. from 1963. In Poland there was a rapid expansion to a peak of 90,000 metric tons in 1965, since when decrease has been equally rapid. Up to 1958, about 50 per cent. of the total catch of the U.S.S.R. was salted, although the proportion was declining. Smoking was the main technique employed in fish preservation in Poland before the Second World War,[5] and the quantity smoked in the U.S.S.R. increased more than 150 per cent. between 1940 and 1946. Smoking and drying, however, now appear to play a minor role in Eastern Europe, and only about 5 per cent of the Russian catch was so treated in 1958. In Western Europe, the production of cured fish has stagnated or fallen. Formerly curing dominated the processing of the main exporters, and as recently as 1958, 85 per cent. of the Faroese catch was cod for salting, although the proportion was only 44 per cent. a decade later. Although the total Norwegian tonnage (of cod and herring) which goes for curing is still large at over 300,000 metric tons, it has been falling and now represents only 12 per cent. of the total catch; the Icelandic tonnage is about one-half that of Norway, but can still be up to 25 per cent. of all landings. The biggest proportion which goes for curing is the 30 per cent of Holland, which has retained a specialized market in cured herring. The Iberian countries have a large proportion of their catches cured, mainly in the cod from the Grand Banks which is salted aboard. In both Spain and Portugal this amounts to 170,000 to 200,000 metric tons per year, and in Portugal 40 per cent. of the total national catch is cured. In most other countries curing is now of minor importance. In Britain, even the market for smoked fish (the highest value section in cured fish) has been notably declining, and for both herring and white fish declined between 60 per cent. and 70 per cent. between 1947 and 1960[6]. On the other hand, smoking of herring, sprat, mackerel and various white fish is the main processing industry of Belgium, and accounts for over 50 per cent. of all processing.[7]

For fish which is used for food, the preservation process that has grown most vigorously in post-war years is that of freezing; it has the advantage of preserving the original quality of the product best, and the frozen sector represents one of high value. Freezing has been developed throughout Europe in some degree: it has been the most rapidly expanding sector in the highly developed countries of the north and west, but it has also been considerably developed in such countries as Spain, Italy and Greece in the Mediterranean zone, where danger of spoilage with higher temperatures is greater. Freezing has also been integral in the planned long-range expansion of the East European fleets. The greater value of frozen storage space has accelerated the trend towards filleting of fish, as the bulk of fillets is less than half that of round fish.

In most countries, freezing is done in plants at fishing bases from high-quality products in fresh landings. In Denmark and Iceland, which are among the leading producers, the fish is caught in near waters and freshness can be guaranteed; this is also partly true in Norway, although an increasing proportion is frozen at sea aboard distant-water trawlers, while in Britain part comes from distant-water catches preserved in ice till landing. Production of frozen fish is particularly important for exporting countries. In Iceland freezing began in 1930, and was stimulated by food scarcities in the Second World War. By 1945 there were 68 plants with an aggregate capacity of 600 metric tons of fish per day, and in 1959 the 84 plants had an aggregate capacity of 1,400 metric tons; these figures show an increase in the scale of plant as well as an increase in the number of units. Freezing also began in Norway in the inter-war period, and the 363 plants in 1962 were a twelve-fold increase over the 1930 total. In Norway, however, trends in the increase in size of plant have been somewhat curbed by the government policy of encouraging decentralized operation, partly to give employment in scattered communities.[8]

Freezing at sea has been most developed by the Eastern European countries and by West Germany, as a consequence of the great distance to the grounds exploited: inevitably the overhead costs of operation at sea are substantially higher. The Russian total production of nearly two million metric tons of frozen fish in 1968, however, came mainly from distant-water fleets, and was a three-fold increase over the 1958 figures. There were 206 freezing plants in the U.S.S.R. in 1957, and at least one-third of the national investment on food-freezing is in fish.[9] In 1965 over one half of the freezing plant capacity of the country was in the European part. The fleets based in the same section of the country had in 1966 the capacity to freeze at sea nearly 10,000 tons per day.[10] A feature of the organization in the U.S.S.R. is the freezing of fish at sea (especially pelagic species) for later canning, smoking and salting at the ports. To a lesser extent Spain and France have developed distant-water freezing for tuna and crustaceans in the Central and South Atlantic. France also produces frozen white fish fillets on distant-water trawlers, and Britain, too, has begun to engage in this.

Canning is another process which gives a high value product. It developed earlier than freezing, and has the advantage of not requiring a specialized cold chain for distribution. The main expansion phase in canning in Western Europe was before the Second World War, and in Norway in the period 1910–18, when it was largely fostered by the stimulus to export of war-time conditions. In Eastern Europe canning has continued to expand vigorously since the Second World War, and the U.S.S.R. is now second in the world only to the U.S.A. in production. The output of 689 million standard cans in 1956 in the U.S.S.R. was more than 250 per cent. greater than that of 1950, and more than 70 times the production in 1913. Canneries have been built on rivers as well as at landing ports, and the total in 1957 was 176; floating canneries are also in operation with the distant-water fleets. Of some 700 fish items on the Russian market, 200 are canned.[11] The rate of expansion has slowed somewhat in more recent years, and in 1965, the total production was

N

960 million standard cans, 62 per cent. of them being produced in the European part of the country.[12] A prominent development of the post war period has been the concentration of the main part of the production in large units each producing over 15 million standard cans per year.

In Western Europe the leading countries in canning are West Germany and France. In the former the annual production of c.135,000 metric tons takes over 20 per cent. of the national catch, although the production has somewhat declined. In France, sardines, tuna and mackerel are canned in quantity, and the total production is c.90,000 metric tons per year, and there was a 50 per cent. expansion between 1958 and 1968. Portugal and Spain both have important canning industries, mainly for sardines, but tuna also feature prominently in Spain. Each produces around 50,000 metric tons per year. In 1955 in Portugal there were 250 canneries with a total labour force of 22,500: processing was labour intensive, as skinning and boning of the fish were done by hand.[13] The average size of plant in Spain is smaller as the 748 canneries in 1961 had a total labour force of 22,826.[14] Fish canning is relatively important in the Netherlands, where about 7 per cent. of the catch, consisting mainly of herring and crustaceans, is canned. The tonnage canned in Norway, Denmark and Britain is considerable, although in each only about 1 per cent. of the catches are so used. Canning has increased little in these countries since the Second World War, although there has been a tendency for it to be concentrated in bigger units, and in Norway the number of canneries halved between 1946 and 1962: at the latter date there were 142.

Processing is very largely located in fishing bases, and with modern tendencies of concentration, the bigger ports tend to dominate processing more than they dominate landings through the gathering of additional supplies by overland delivery. This is instanced by Hamburg—Altona in West Germany, Göteborg in Sweden, Ijmuiden in Holland, and Hull, Grimsby and Aberdeen in Britain. Some specialization does occur in different ports, as is instanced by the building up of Kaliningrad as the main herring port of the U.S.S.R., and Matozinhos (the chief Portuguese sardine port) has

70 canneries. In Western Europe, there has been a limited recent decentralization of fish processing, mainly of freezing, where it is an important advantage to secure freshly caught supplies, especially for shell fish. In Eastern Europe, the main development appears to have been that of big integrated 'combines'. These are located mainly at the main landing ports (including Murmansk, Kaliningrad, Rostock and Swinoujsie), but there is also a large one at Astrakhan on the Caspian Sea, and there are also combines at inland cities like Moscow and Kiev. The Murmansk combine, for example, covers a site of 100 acres, and has sections for filleting, freezing, salting, smoking, reduction to meal and oil, and the making of cans and of ice. It has a labour force of 5,500, and there has been a considerable development of automatic unloading and processing equipment.

Very few statistics are available for the labour forces involved in processing in Europe. It would appear that in most cases the expansion of modern processing methods like filleting, freezing and reduction have largely compensated for the run-down in employment in more traditional methods like salting and drying: in Norway, for example, the number of fish workers has been fairly stable since 1946 at around 14,000. In Denmark the post-war expansion in reduction and freezing has generated additional employment, but in Scotland the continued run-down in herring curing—a labour intensive activity which formerly dominated processing—has led to the halving of the labour force from the inter-war period, although it now appears to have stabilized. The major problem in Western Europe is becoming that of retaining labour, as employment in fish processing is often left if an alternative becomes available. In Eastern Europe, employment has grown; the scale of expansion is not clear, but in the U.S.S.R. the number of engineers and technicians working in fish processing stood at 24,000 in 1959, which was a 270 per cent. increase over 1940. A big proportion of the 330,000 not aboard ship in the fishing industry in 1960 must have been employed in fish processing.

The effect of processing on board in reducing the weight of fish brought back to landing bases is evident for most countries in the differences between catch and landed weights published for most countries in the F.A.O. statistics. Unfortunately, only catch weights are available for most of the Eastern European countries which have developed processing at sea to the greatest degree. Many varieties are not processed to any great extent on board, but figures are often available for different species, and these indicate the sectors in which processing aboard is important. When catches are gutted only, the weight is usually reduced by between 10 per cent. and 15 per cent. On the other hand, splitting and salting of white fish aboard reduces the weight by up to 60 per cent., although pelagic species are usually salted in brine, and the weight reduction is not much more than is achieved by gutting. Filleting is most effective in reducing weight, as it produces 30 per cent. to 50 per cent. the weight of round fish.

In countries where the main fisheries are inshore, nearly all processing is done on land: in Denmark and Italy, for example, gutting of some species aboard only reduces the total caught weight by some 2 per cent. Although Dutch fisheries are prosecuted almost completely in the North Sea, the concentration on herring, which are often processed aboard, leads to a reduction of 13 per cent. Norway and Iceland are leading fishing nations, but the concentration of operations in home waters is associated with a limited amount of processing on board, and landed weights are only 7 per cent. less than catch weights. This is achieved mainly in the important cod group, where salting or filleting before landing reduces the weight by 25 per cent. It is the countries which are most involved in distant-water operation that have greatest differences between caught and landed weights, especially when they are involved in salting or in freezing fillets. In Portugal, the difference between total catch and landed weights is 28 per cent., due largely to the big catches of cod salted aboard dory-schooners on the Grand Banks: for the cod group the difference is actually 63 per cent. For the same group in France and Spain it is 45 per cent.; and in these two countries there is a 20 per cent. difference between total catch and landed weights, due partly to the gutting aboard of species like tuna, sharks and hake. Most processing before landing is actually achieved in the Faroe Islands, where there is a 32 per cent. difference, largely due to the proportion of salt cod: formerly this was mainly dry salted at shore bases in Greenland, and the difference was over 50 per cent. until 1958. West Germany has most extensively developed the techniques of filleting and freezing aboard in Western Europe, and also salts part of its herring on the catchers: the overall reduction of weight is 27 per cent., although in the cod group it is 40 per cent. and in herring 25 per cent. In countries like Britain and Belgium, which have distant-water fleets but on which there is little filleting, the difference between catch and landed weights is around 10 per cent. The only country in Eastern Europe for which separate catch and landed weights are published is Roumania: here the expansion of off-shore fisheries in the Black Sea leads to a difference of 12 per cent.

The most vigorous development of processing at sea has undoubtedly been in Eastern Europe, particularly in the U.S.S.R. In 1965, when the total catch was 5,100,000 metric tons, 75 per cent. of the processing was done at sea.[15] There has been planned expansion of fleets of factory-trawlers and mother-ships; some of the latter are vessels of over 10,000 tons. In 1959 there were at least 24 2,500-ton 'Pushkin' factory trawlers, and at the same time 70 of 3,500 tons were being built. At the same date the total number of freezing units in the fleet was 300. Aggregate freezing capacity afloat was given as 400,000 metric tons in 1959, and had increased eight-fold in the previous decade: it was planned to expand to 900,000 metric tons by 1965.[16]

ALLOCATION OF FISH TO DIFFERENT PROCESSING USES: *Examples from Norway & Britain*

Trends in processing may be illustrated from Norway and Britain: Figure 21 shows the volume of production of the more important Norwegian products between 1959 and 1966, and also the changing utilization of the British herring catch over the post-war period: this sector of the British catch is subjected to a bigger variety of processes than any other.

In Norway, the outstanding trend since before the Second World War is the rise in reduction products (meal and oil) from herring; also prominent is the decrease in products

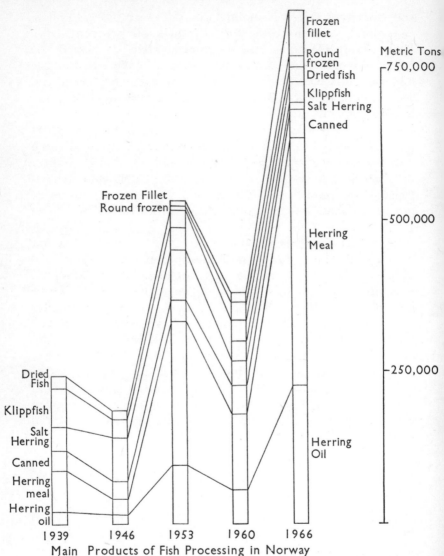

Main Products of Fish Processing in Norway

Figure 21(a). Trends in processing uses in Norway.

preserved by traditional methods (dried fish, klippfish and salt herring) and the increase in the tonnage that is frozen. The quantity processed has fluctuated, mainly as a result of fluctuations in the herring catch, but the general trend has

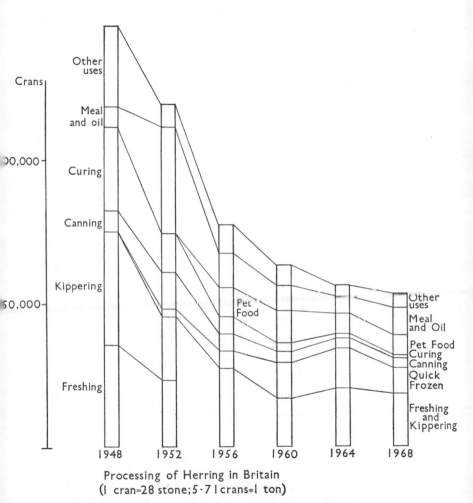

Processing of Herring in Britain
(I cran=28 stone; 5·7 I crans=I ton)

Figure 21(b). Trends in processing uses of herring in Britain.

been upward, and the total quantity processed in 1966 was more than 250 per cent. greater than in 1939. In 1946 the tonnage allocated to the higher value use in salting was greater than that of meal and oil produced in reduction. Although canning has expanded slightly, the main high-value products now are frozen fish, and the output of frozen fillets has grown more rapidly than that of the lower-value frozen round fish. Despite the growth of the frozen sector, dried and klippfish remain an important sector in Norwegian processing, and more fish are still used for these processes although the weight loss is greater.

Figure 21 also shows the decrease and changing structure of the market for herring in Britain during the 1948–68 period, in which landings fell by 64 per cent. In Britain uses for reduction and also for pet food are still largely surplus outlets; there is a big price differential between those used for human consumption and other uses, and there has been limited incentive for fleets to fish specifically for reduction and pet food. Prominent also is the run-down in the once dominant curing industry, which before the First World War absorbed over 500,000 crans per year. Utilization for fresh herring and kippering has also declined, although these have been consistently the main market sector since 1948. Canning has declined somewhat less, and the main increase has been in the proportion used for quick-freezing. Meal and oil utilization has fluctuated, as would be expected for the sector of which the main function has been to absorb supplies surplus to other uses.

REFERENCES

1   O. Onarheim, 'The Development of Factory Ships and of Related Media in Fisheries', in *Some Aspects of Fisheries Economics*, ed. G. M. Gerhardsen, 1964, I, p. 89.

2   E.E.C., *Rapport Sur la Situation du Secteur de la Pêche*, 1966, p. 168.

3   G. M. Gerhardsen, *Fiskeriene i Norge*, 1964, p. 173.

4   O. Bjornsson, 'The Icelandic Fisheries', in *Atlantic Ocean Fisheries*, ed. G. Borgstrom and A. J. Heighway, 1961, p. 263.

5   Z. Fruczek, E. Kordyl, and S. Laszczynski, 'Development and Present State of Polish Fisheries', in *Atlantic Ocean Fisheries*, p. 92.

6   C. L. Cutting, 'The Fishing Industry of Great Britain: Handling and Marketing', in *Atlantic Ocean Fisheries*, p. 131.

7   S. Lefevere, 'Fish Processing in Belgium', in *Atlantic Ocean Fisheries*, p. 140.

8   G. M. Gerhardsen, *Ibid.*, p. 161.

9   M. M. Ovchynnyck, 'Development of Some Marine and Inland Russian Fisheries and Fish Utilisation', in *Atlantic Ocean Fisheries*, pp. 273, 274.

10   N. P. Sisoev, *Ekonomika Ribnoi Promyshlennosti*, p. 206.

11   M. M. Ovchynnyck, *Ibid.*, p. 274.

12   N. P. Sisoev, *Ibid.*, p. 209.

13   R. I. Houk, 'The Portuguese Fishing Industry', in *Atlantic Ocean Fisheries*, p. 171.

14   *Economic and Social Development Program for Spain*, 1964–67, p. 173.

15   E. D. Kustov, *Geografiya Ribnoi Promyshlennosti*, 1968, p. 18.

16   G. Borgstrom, 'The Atlantic Ocean Fisheries of the U.S.S.R.', in *Atlantic Ocean Fisheries*, p. 302.

CHAPTER X

The International Fish Trade in Europe

Trade in fish is related to areal variations in supply and demand on a series of scales from the continental to the local. The composition of the trade is related to the structure of supply and of demand, and distance between points of supply and of demand is a main factor in determining trade flows. Much of the present pattern of trade is of modern growth, and is basically related to the extensive deployment of railway and steamship transport in the decades around 1900, although now road transport plays an important role, and an increasing proportion of the more valuable products is conveyed by air.

The pattern of movement is now complex, and trade in fish and fish products is part of the general system of expanding trade in the sophisticated economies of a highly developed continent. All countries both import and export fish, although the scale of trade varies greatly, as does the relative importance of the trade to the different nations and regions. In a system in which competition is growing, the volume of trade is primarily determined by the established needs and purchasing power in particular markets. The main part of the systems that supply these markets operates within individual countries, but there are increasingly important international connections between supply systems. In this context the modern formation of supra-national groups is especially noteworthy: these include the areally continuous six countries of the E.E.C. (European Economic Community), the seven in the E.F.T.A. group (European Free Trade Area), which are distributed around the E.E.C. group, and in Eastern Europe the COMECON group.

The fractions of the fish catches of the European nations

which enter into trade cannot be stated with any precision, because of the loss of weight in processing which varies for different species and methods: this loss is especially heavy in fish subjected to drying and reduction. Even for major exporters like Iceland and Norway, the export tonnage is less than half the landed tonnage, and for most countries it is a much smaller fraction. It is noteworthy, however, that because of the value added by processing, the export value in Iceland and Norway is between 50 per cent. and 100 per cent. of the landed value.

TRADE BETWEEN EUROPE & OTHER CONTINENTS

On the global scale, fish features in the trade of Europe with all other continents, although for the most part on a minor scale; and its importance in the total patterns of inter-continental trade is in most cases slight.

Europe is now a net importer of fish and fish products, because of the great increase in the import of fishmeal and fish oil since the mid-1950s, and which is now by far the greatest trade item by tonnage. On the world scale, meal and oil now represent over one-half the total trade in fish and fish products; in 1968 the total world trade in reduction products was nearly 4,500,000 metric tons. European nations now account for nearly 60 per cent. of the imports of meal and oil, two-thirds of the European import coming from Peru with her great anchovetta fisheries for reduction; and Chile and South Africa are both big exporters to Europe. This commerce in the products of industrial fisheries, which dominates the inward trade in fish products to Europe, is of low unit value (at around U.S. $100 per metric ton) and is transported in bulk. Most important among other European imports are high-value products which can be economically carried in small quantities. These include canned and frozen salmon from Canada, and canned salmon, shrimps and crabs from Japan, and the unit value of such products ranges from 15 to 20 times that of fishmeal and oil. The import of fish for food, other than for the luxury market, to Europe is of minor importance. Formerly the trade in salt cod from Newfoundland to the Mediterranean countries helped provide a food staple,

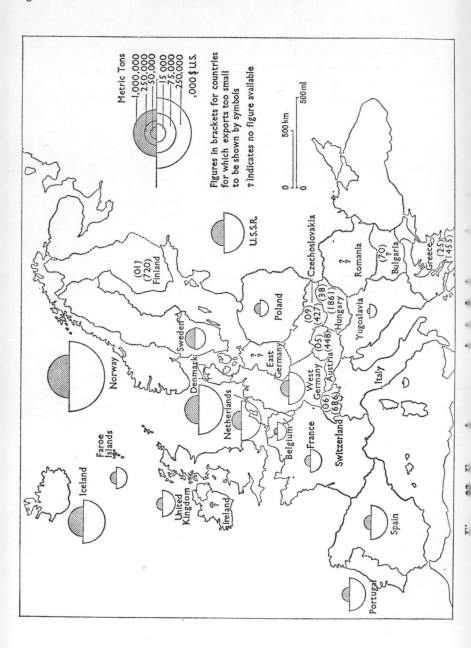

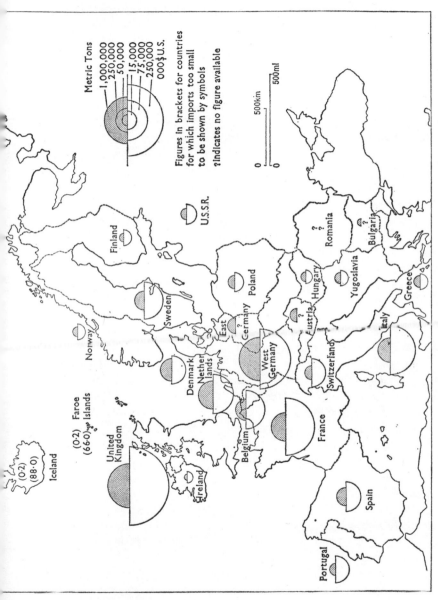

Figure 23. Imports by tonnage and value for the countries of Europe in 1967.

but this is now dominated by producers within Europe. An item which has come more recently into the European markets is tuna from Japan, especially to Italy and West Germany.

Exports of fish from Europe to other continents also move in channels for higher and lower value commodities, although the value differential here is considerably less wide. The highest-value exports from Europe are various shell-fish which are preserved by freezing or canning, and the general level of the value of which is around U.S. $2,000/metric ton. Within the lower bracket come such products as dried cod and salt herring, together with frozen items like fillets of cod and halibut, and the general value of this group is in the range of 10 per cent. to 20 per cent. that of shell-fish. Within this group, however, there is an effective value differential, as dried cod are a relatively concentrated food, as up to 80 per cent. of the weight may be lost in drying, while fillets give from 30 per cent. to 50 per cent. of the round fish.[1]

The higher value products are dispatched mainly to the U.S.A. which now imports almost one-half the fish it consumes, and the import to the U.S.A. of frozen fillets and shell-fish, and canned sardines, anchovy and herring, has been increasing rapidly. The U.S.A. catch has stagnated since before the Second World War, and part of the reason for this is that production costs are considerably higher there in a food-producing industry which is still basically at a hunting level, and in which it is particularly difficult to approach the levels of productivity found in other sectors of the economy of the U.S.A. Alternative foodstuffs also tend to be over-produced, and the U.S. government has not seen fit to promote its own fisheries in the way that most of the European nations have done. Of lesser importance for high-value products from Europe are markets in such developed countries as Australia, New Zealand and South Africa, which have large populations of European origin.

The lower-value sector of exports from the continent of Europe is largely directed to developing countries, and the most important item consists of cod preserved by the more traditional methods of drying and salting. Here Brazil is a market for the salt 'klippfish', and Norway, Iceland, the Faroe Islands and France compete in providing supplies. Norway has also built up this century an important market for air-dried 'stockfish' in West Africa: the main recipient in this trade is Nigeria, and the trade was seriously disrupted by the Biafran War. Although the U.S.A. is Europe's main overseas market for high-value products, it also is an importer of dried cod and salt herring, largely to provide for such immigrant minorities in its population as the Puerto Ricans. The former export of salt herring, especially from Britain, to the Caribbean, has now dwindled to negligible proportions. Other products which are a minor component of export lists from Europe are smoked items—mainly herring—which are sent to a variety of destinations in the Americas, Africa and Australasia.

A feature of a part of the overseas trade in fish from Europe is that it is linked with political affiliations. Thus exports from Portugal include the high-value products of sardines to her overseas African territories of Angola and Mozambique; France sends dried cod to her Caribbean island territories of Guadeloupe and Martinique; and the U.S.S.R. has in recent years built up a considerable export of salt fish to Cuba, and of frozen products to Cuba, Togo and the United Arab Republic.

INTERNAL EUROPEAN TRADE

The major part of the European fish trade, by value if not by volume, is internal, and Figures 22 and 23 show imports and exports for each country by tonnage. The trade moves along two main directions. The major direction is peripheral, with the exchange of commodities among the seaboard nations which are also producers, but also important is movement away from the coastlands into the interior to supply countries and regions for which home supplies are restricted to inland waters. The volumes of trade flows decrease away from the open ocean, and (less prominently) away from the Baltic and Mediterranean Seas.

Part of this movement goes across international frontiers by land; this includes the export of salt herring from the Netherlands to West Germany, the movement of salt cod from West Germany to France, Belgium and Italy, the interchange of fresh and chilled fish between France and Belgium, and the export of frozen mussels from the Netherlands to France.

Of bigger volume is the seaborne movement between countries, which includes the outward trade of the major fish exporters of Norway, Iceland, the Faroe Islands and (in part) of Denmark. Included is the export of chilled and frozen fish and fillets from these countries to industrial nations like Britain and West Germany, and the export of fishmeal and oil to these and a variety of other nations throughout the continent. The export of Portuguese canned sardines is another major item which goes very largely by sea.

An example of the costs of exporting by sea is available from the Faroe Islands for 1967, and this shows the effect of distance on costs.[2] From these islands, the average cost of sending salt cod to Norway was c. 2½ per cent. of the product value, to Denmark c. 3 per cent., to the north coast of Spain c. 4 per cent., to Italy c. 6 per cent. and to Greece c. 7½ per cent. These indicated a tapering range of freights with distance, which ranges from c. 400 miles to Norway to c. 3,700 miles to Greece. When export takes place in refrigerated holds in summer to Italy and Greece, the proportion rises to c. 8½ per cent. and c. 10 per cent. respectively. The product sent to European markets is mainly wet-salted, but the more valuable dry-salted cod, which is c. 75 per cent. more valuable than wet-salted, is better able to bear transport costs: the distance to Brazil is over 6,000 miles, but freight rates are c. 6 per cent. of the product value by ordinary ship, and c. 8½ per cent. by refrigerated vessel. Salt fish is a relatively concentrated foodstuff, and has relatively high value in relation to its bulk. It is significant that the transport costs on frozen fillets from the Faroe Islands to the U.S.A. were c. 8½ per cent. although the distance is c. 3,000 miles.

For the most valuable products—shell-fish—air transport is feasible to send them fresh to markets over the longer distances. Air freight on lobsters from the Western and Northern Isles of Scotland to Paris, for example, ranges from 10 per cent. to 25 per cent. of the product value, according to the size of the consignment and the season.

While distance is a factor in conditioning the dimensions of trade flows, the regulations imposed by nations in the form of import tariffs and quotas are generally more important. Thus countries with big volumes of consumption like Britain and Germany have imposed dues on incoming fish which give measures of protection and stability to their own fishing industries against competition from main exporters like Norway, Denmark, Iceland and the Faroe Islands. The imposition of quotas to the same end is instanced in Britain's trading arrangements with the same countries.

In the 1950s, levies on imported fish in Europe were generally in the range of 15 per cent. to 50 per cent.[3], and fresh fish especially could come in only over high tariff walls. There has been some subsequent modifications of these rates, but this has mainly been within the different economic blocks. Within the E.E.C., tariffs on fish and fish products were reduced by 55 per cent. in 1965, and have since been phased out; but at the same time the E.E.C. aims at a common external tariff of 15 per cent. to 25 per cent. on fish and fish products.[4] A notable effect of lower tariffs has been the increase in the export of fresh and chilled fish from Belgium to France, which increased two-and-one-half times between 1966 and 1968. Reduction of duties on fish in the E.F.T.A. group led to their being phased out in 1970, but there are special provisions in the treaty to guard against 'dumping' of fish, as well as farm produce, in the markets of member countries. Fish exports are important in the economies of Norway, Portugal and Denmark, although most E.F.T.A. exports are to countries outside the group, and Britain takes two-thirds of all E.F.T.A. imports. The most prominent effect of reduction of tariffs within the group has been the increase in the export of frozen cod and haddock fillets to Britain from Norway, which nearly doubled between 1966 and 1968. It is intended that some products (including canned fish, fishmeal and frozen fillets) will be classed as industrial goods and trade in them will be completely unrestricted; but this will not apply to fresh, chilled, salted and dried fish, nor to crustaceans and molluscs. Formerly much of the fish consumed in the COMECON countries was bought from Western Europe, but linked with the vigorous growth of fisheries in COMECON has been the development of trade among its members to the point that imports from Western Europe are of slight importance. In Western Europe, it is not clear how far the formation of trading blocks will modify the pattern of the fish trade, even in the long term. The external trade of the E.E.C. is more than twice the internal, Denmark maintains its position as the main exporter to E.E.C. countries, and Japan continues to be important. Iceland has trade agreements with the East European countries, and is now their main trading partner in Western Europe, salt herring to Poland and frozen fillets to the U.S.S.R. being important exports; and the Faroe Islands send salt herring to Czecho-slovakia in COMECON. While much of the fish trade is now part of a multilateral pattern of European trade, difficulties can be encountered in bilateral trade agreements which are characteristic of trade between nations in Eastern and Western Europe. The post-war decline of Norwegian exports to Eastern Europe is related to the difficulty in getting acceptable products in exchange, as well as to the post-war rise in catches in Eastern Europe itself.[5]

Trends in the volume of the fish trade in Europe have been markedly upward during the present century, although in the period from 1938, during which comprehensive figures are available, the fraction of Europe's own catches which have been traded has remained substantially constant at around 22 per cent. This indicates that internal trade has increased *pari passu* with external, apart from the post-war import of reduction products from other continents. The big increase in total European catch, together with the stagnating consumption rates for most fish products in industrial nations, has meant that the modern international fish market tends to be a buyer's rather than a seller's one.

Between 1948 and 1968 the gross imports of fish and fish products of the European nations (excluding the U.S.S.R.) rose from 1,408,000 to 4,802,000 metric tons, or by over 240 per cent.; in the same period, exports rose from 1,243,000 to 2,675,000 metric tons, or by over 110 per cent. The value of imports in 1968 was U.S. $1,258,000,000, or 200 per cent. above 1948, although when allowance is made for inflation the increase in real value of imports was only around one-half this figure: this big decrease in real value per metric ton is due to the large role now played by the products of reduction. The increase in value of exports to U.S. $815,000,000 represented a rise of c.140 per cent. by value; and the rise in real value per metric ton of exports relative to imports is due to the much bigger inward trade in reduction products. A feature of the trade now is that for almost every country the average value per ton of exports is higher than for imports, despite the costs added to the latter for transport. The reason for this is again the big volumes of reduction products in imports, and this comprises a big fraction of imports even in a country like Norway that has itself a big export of meal and oil. The only country for which the value per ton of imports exceeds that of exports is Sweden, where 'skrapfisk' for reduction in Denmark is a major item on the outward trade.

Within the gross figures, and also in the U.S.S.R., there have been considerable changes in the composition of the European fish trade. Over the post-war period almost every country has increased its imports of fish; the only exceptions are Iceland, which is virtually self-sufficient, and the U.S.S.R., which between 1958 and 1968 reduced its imports from 139,000 to 41,000 metric tons. This was largely due to the build-up of its own fleets to supply products which were formerly bought from other nations; and with the increase of the fleets of East Germany, its imports have fluctuated without any long-term rise.

The scale of the increase in imports in other countries has varied considerably. The lowest increases have been shown by long-established fishing countries, and by those which also have made big increases in their own catching efforts. Among the former may be cited Britain, where the increase in tonnage between 1948 and 1968 was of the order of 130 per cent., and Portugal, where there was little rise till after 1965; among the latter Greece, where the growth was less than double, is an example. Britain, with West Germany, still leads the lists of importing nations with between 950,000 and 1,000,000 metric tons per year. At the other end of the scale, land-locked countries like Switzerland, Austria and Czechoslovakia have bought increasing amounts from coastal nations, and in the first mentioned the import tonnage has actually increased fourteen times over the twenty-year period. While there has been a persistent post-war rise in established fishing countries like Spain, West Germany and France, in others, like Norway and Yugoslavia, recent years have shown a fluctuating level of imports, suggesting the tendency to an equilibrium in the import sector of their trade.

There has been a rising trend of fish exports in most countries, although the overall rise is less marked than for imports. In the post-war period there have been several changes in relative importance among the leading exporting nations. These include the rise to the front rank in importance of Denmark and the U.S.S.R., which both now export well over 300,000 metric tons annually. In addition, the Netherlands and Sweden now export about 200,000 metric tons annually, while the rise in Spain and Portugal has put their figures at around 100,000 metric tons, on a par with those of West Germany. In the longer established exporters, Norway and Iceland, exports have risen more than double in the post-war years (though less consistently); these are both still in the front rank of exporters, and Norway in 1967 exported over 940,000 metric tons, while in 1966 the figures for Iceland were nearly 500,000 metric tons. A fluctuating rise has also characterized the exports of the Faroe Islands, where the exports per capita are on a par with those of Iceland.

The most spectacular rates of increase in export trade have largely been realized in the centrally planned economies of Eastern Europe. Although no figures are available for East Germany, the exports of both the U.S.S.R. and Poland increased by over ten times in the decade between 1958 and 1968. In Western Europe, Spain is the only nation with a comparable rate of expansion. Spanish exports increased fourteen times between 1948 and 1968, and four times between 1958 and 1968; at the latter date the figures approached 100,000 metric tons. It is notable that there have been relatively big increases in the export trade in fish even from countries which are heavy net importers; these include countries in central and southeast Europe. Thus although Italy's exports in 1968 were only 12,000 metric tons, they were 17 times those of 1948, and in recent years the exports of Switzerland and Austria have been running at five to eight times those of 1948. This reflects mainly a widening international market for high-value products like river fish and molluscs.

Great Britain is a lone example of a country in which the

export trade has been decreasing, largely as a result of the continued decline in overseas markets for cured herring. Between the end of the Second World War and the early 1960s, the export trade decreased by half to around 50,000 metric tons, but there has been a more recent resurgence.

There are wide differences in the values for the trade of different countries. In exports, the gross annual value of the Norwegian trade now runs at over U.S. $200,000,000, and that of Denmark is well over U.S. $100,000,000, while the Icelandic total fluctuates around this latter figure. At the other end of the scale, exports from land-locked countries like Austria and Switzerland are valued at well under U.S. $1,000,000. Britain is the leading importing nation by value, with over U.S. $250,000,000, and West Germany and France are in the range between U.S. $150,000,000 and U.S. $200,000,000. At the other end of the scale, Icelandic imports are considerably less than U.S. $100,000.

The countries which are the leading exporters by gross tonnage and value have lower average values per ton for exports than most countries, as the more valuable products constitute a smaller portion of their outward trade. The southern nations of Spain, Portugal, France and Italy all have an average value for exports of $475 per metric ton or more, while those of Norway, Iceland, Denmark and the Netherlands are under $350 per metric ton, and of Sweden under $150. The highest unit value of exports is that from the landlocked countries of Austria and Switzerland, which consist of the more valuable fresh water products, and their value is in the range of $800 to $1,000 per metric ton. The average unit value of imports is reduced now by the big volume of reduction products, and only those of France, Italy, Portugal, Sweden, Switzerland and Britain are in the region of $400 per metric ton, while those of Denmark, Poland and Yugoslavia are under $200 per metric ton.

The composition of the trade flows in fish is more determined by distance than are the volumes, as the perishability of the product necessitates greater expenditure on processing for items carried over longer distances. Some shell-fish, which are particularly susceptible to rapid deterioration, must be delivered alive to markets or processing plants: these include lobster and crab. No measures are generally necessary for short-distance delivery of fresh fish for immediate consumption. For intermediate distances, preservation on ice is often sufficient, especially if transport is rapid. The longest distances demand more sophisticated processing methods and storage facilities and are consequently more costly: these include salting, smoking, drying, canning and freezing, and reduction processing can be added. While the statistics gathered by F.A.O. can in part show the trade flows, it is unfortunate for this purpose that fresh, chilled and frozen constitute one of the seven major groups for which statistics are presented, and only in part is there a break-down available.

At the international level, there is very limited trade in fresh fish, other than on ice. The modern trade in fish very largely grew in iced fish, and this sector still operates mainly within national frontiers, although it is now an important part of international trade. Included now is the export of iced cod and other fish from Iceland to Britain and West Germany, of salmon from Denmark to Belgium, and of herring from Norway and Britain to Germany. There is some seasonal variation in the trade in iced fish, as higher summer temperatures mean quicker melting, and it is general practice to adjust the ratio of ice to fish.

Products preserved by the various processing methods have greater freedom of movement, and the longer distance trade consists almost wholly of such products, while they comprise an important component of the trade within the coastal zone also. Prominent now are the greater efforts being made in the majority of countries to increase trade in the more valuable processed items like frozen and canned shell-fish and fillets. The main details of the fish trade within Europe can be shown by considering each of the seven classes of items recognized by F.A.O.

Fresh, Chilled & Frozen Fish. This is the biggest sector by both tonnage and value in the trade, apart from that in reduction products. The location of the producing country is here an important—or even decisive—factor in its trade. The leading exporter is Denmark, which has big markets in West Germany, Britain, France and Sweden, all within 300 miles; the total Danish export in this sector now runs at over 200,000 metric tons per year, and its value at over U.S. $80,000,000. The Netherlands is also a main exporter, and is even closer to its main markets in Belgium, West Germany and northern France. Norway and Iceland are also major exporters in this category, despite their remoteness from main markets. The tonnage of Swedish exports is second in Western Europe only to the Danish, but its value is less than one-quarter as they consist mainly of herring for reduction. The U.S.S.R. is now the greatest single exporter of frozen products, the annual tonnage of which approaches 200,000 metric tons; much of this trade is with COMECON countries, but also with countries in other continents such as Cuba and Togo; and the greatest single importer is Japan which absorbs more than one-quarter of the total, which comes from catches landed on the Pacific seaboard of the U.S.S.R. These frozen products are sold at values much below those of Western Europe, the average value in 1968 being U.S. $135 per metric ton, compared with western prices of c. US. $600 per metric ton for frozen white fish fillets, and c. U.S. $180 for frozen herring. With its expanding catches, the U.S.S.R. could be a strong competitor in the future in one of the highest-value parts of the European markets. The leading importers in this sector are the big industrial nations of West Germany, France and Britain, for which the inward trade is over 150,000 metric tons per year, and the value over U.S. $50,000,000; and this sector is now relatively important for all European nations.

Dried Salted & Smoked Fish. The main part of this sector is the export from the northern producers, Norway, Iceland and the Faroe Islands, which participate in the salt cod trade (page 206), although Spain is now a strong competitor with an export second only to that of Norway. Norway exports a total of c. 80,000 metric tons per year in this sector with a gross value of c. U.S. $50,000,000. The leading importers are the Mediterranean nations of Italy, Spain, Portugal and Greece. Italy, which (in contrast to Spain and Portugal) produces little salt cod of her own, imports over 50,000 metric tons per year, which is more than twice that of any other European nation. The other main component in this trade is that in salt cured herring, for which the Netherlands and Iceland are the main exporters, and West Germany the main importer in Western Europe. There are no gross figures available for the Eastern European nations, which are the main consumers of cured herring.

Crustaceans. Although the total European trade in shell-fish scarcely exceeds 150,000 metric tons, the value is relatively high. The main exporters are Holland (c. 66,000 metric tons in 1968) and Spain (c. 22,000 metric tons in 1968), the former consisting mainly of cultivated mussels and the latter of oysters. The value of these, however, is only c. U.S. $200 per metric ton, compared with the export of lobster, nephrops and scallops fom Britain and Norway at average values between U.S. $1,500 and $2,000 per metric ton. By far the biggest importer is France, which alone accounts for more than one-half the total import of crustaceans in Europe.

Fish Products & Preparations, Canned or Otherwise. This group comprises mainly canned fish, but with the addition of a variety of specialized herring cures from Iceland and Norway, together with a minor amount of fish roes and fish pastes. Europe is a net importer in this sector, gross imports being around 300,000 metric tons and exports 200,000 metric tons. The import from outside the continent is composed largely of salmon from Canada and Japan, and tuna from the latter. Within Europe, the outstanding exporter is Portugal with over 60,000 metric tons consisting very largely of sardines, and Norway's canned herring and sprat exports are around one-third this amount. The main importing countries are in Northern Europe, with Britain, West Germany and France heading the lists; the trade also extends into Central Europe, and Austria and Switzerland are leading importers relative to their population. The most valuable fish product of all is in this sector—that of caviar from the U.S.S.R.; the value is over U.S. $4,000 per metric ton, although the annual export is little over 1,000 metric tons.

Crustacean & Mollusc Preparations, Canned or Otherwise. This sector by volume is very small, but by unit value is the highest, the average approaching U.S. $2,000 per metric ton. Total

European exports at c. 14,000 metric tons are headed by the U.S.S.R., Spain and the Netherlands, although imports are 150 per cent. greater. France is the outstanding importer, followed by Britain, Sweden and Belgium.

Oils & Fats. Europe has a big net deficit here, gross exports being only about one-third of the gross imports of c. 800,000 metric tons (1968). At c. U.S. $200 per metric ton, this is the more valuable section of reduction, and the main exporters in Europe are Norway, Denmark and Iceland as a result of their 'industrial' fisheries for herring, capelin and sand eels. The main importers are Britain, West Germany and the Netherlands which between them take nearly 600,000 metric tons as a raw material for food and other industries. There is a very limited import of fish oils into the Mediterranean world, partly because of the competition of vegetable oils produced in the countries themselves. Shortage of supplies in Europe in recent years has accentuated the dependence on South American suppliers, and the Netherlands and Germany buy big proportions there, though Britain imports from Norway, Denmark and South Africa.

Fishmeal. The gross European figure for imports at 2,333,000 metric tons (1968) is again three times the export. The Norwegian export exceeds 400,000 metric tons, and the Danish 200,000 metric tons, while Iceland in some years has approached the latter figure. With the now universal employment of fishmeal as an animal feeding stuff, it is more widely distributed throughout Europe than fish oil, but West Germany and Britain are again the main importers with over 500,000 metric tons each.

EXAMPLES OF THE TRADE IN PARTICULAR FISH & THEIR PRODUCTS

Within the broader picture of the fish trade in Europe already discussed, it gives a useful additional perspective to review the trade in particular items from the catching nations through to those in which they are consumed. The most extensively traded items in Europe are those of cod, herring and their products.

The Cod Trade

Cod is the most important single food fish in Europe, and the F.A.O. statistics allow an examination of the main part of the trade at the international level, although some of the details—including most of those for Eastern Europe—are not clear. Although the catches of several of the main producers, including the U.S.S.R., Britain, West Germany, France, Spain and Portugal go substantially to satisfy internal markets, there are big volumes which enter international trade, and which move mainly in a north-west to south-east direction. These are supplemented by what is now a minor amount entering the continent from the east of Canada—especially from Newfoundland; and a minor proportion of the export trade of the European nations moves outside the continent, mainly to the U.S.A., Brazil and Nigeria. The trade has several components, as in addition to the long-standing items of (air-dried) stockfish and wet and dry salted cod, it is also traded as whole fresh and fresh fillet, as well as frozen (generally filleted); in addition cod-liver oil is traded internationally, although its importance is declining in face of competition from animal and synthetic oils.

The main item in the trade is still salt cod, although the frozen fillet sector is now a rival, and indeed is the most important item in the markets of higher purchasing power. Stock-fish and fresh cod also enter prominently into the trade flows, although in these items trade takes more specialized directions.

Salt cod is exported mainly from the northern producers— i.e. Norway, Iceland and the Faroe Islands—to the Mediterranean countries, although France and West Germany are significant exporters also. Of the main importing nations in Europe, the trade into the Iberian nations goes to supplement a big home production but in Italy and Greece provides almost the total supply. Markets outside Europe are relatively minor but include Brazil and the U.S.A.

Included in the trade is that in dry and wet salted fish: the former, which is of higher value, amounted to c. 30,000 metric tons in 1965 and comes mainly from Norway. Brazil, Portugal and Italy are the main importers. Iceland and the

Faroe Islands formerly competed with Norway in the dry salted trade, but have in recent times gone more to the wet salted, which can be made aboard boat. The total European wet salted trade amounted to c. 85,000 metric tons in 1965. Iceland and the Faroe Islands now vie to be the main producers in this sector, but have competition in export from France, West Germany and Norway. The main importers are in Southern Europe, and Italy is outstanding.

In the frozen fillet trade, the leading exporting nations, Norway and Iceland, are rivalled by Germany and Denmark, who also have modernized their processing, and who have more convenient and immediate access to markets. The total trade in 1965 exceeded 100,000 metric tons. In this sector, the U.S.A. is the most important single market for the products of the European nations, and within Europe most of the more developed countries of the north are good customers: these include Britain, Sweden, France, the Netherlands, and (for Icelandic exports) the U.S.S.R.

In the fresh trade, there is a premium on expeditious handling, and virtually all cod in this category are shipped on ice. It is possible, especially during the colder half of the year, to send quality products in this way from the remoter producers of Iceland and the Faroe Islands to continental destinations, and these in fact overshadow those of the other main nations involved, West Germany and Denmark. Precise figures are not available for this sector of the trade, but it appears to have been in the vicinity of 40,000 metric tons in 1965. The main lines of movement in this sector are those from Iceland and Faroe to Britain, but all the coastal nations of the northwest participate to some extent.

Stockfish comes only from Norway and Iceland, and only in these peripheral northern countries are conditions cool enough for its preparation. The total trade in 1965 was 38,000 metric tons, and the main part of this went to Nigeria, while the only significant market in Europe is in Italy.

Cod-liver oil production is now minor, and only Iceland and Norway are involved on any scale. It is traded widely, if in small quantities, and the main markets in Europe in 1965 were Sweden and Denmark.

The Trade in Herring & its Products

Although now exceeded in importance by cod in the fish trade, herring have played an exceptional role in the building of the European fisheries (Chapter IV) and until about the Second World War exceeded cod in their importance in trade. Although the formerly pre-eminent trade in cured herring has dwindled to a shadow of its previous importance, there is now an even greater trade (by tonnage) in herring oil and meal, and international commerce features herring in fresh, frozen, filleted, smoked and canned forms as well as cured.

Iceland and Norway are now outstanding as exporters of herring and herring products, but Denmark, the Netherlands and Sweden are competitors, especially in the sectors involving herring for food.

Trade in the very perishable fresh herring is restricted to relatively short distances, and the bulk of it consists of exchanges between coastal countries with their near neighbours, such as those sent from Denmark to West Germany, or those from the Netherlands to Belgium and France. The biggest tonnage in this sector, however, consists of exports from Sweden to Denmark; they consist mainly of landings by Swedish craft in Danish ports to supply the Danish reduction industry, and there are also some landings by the Faroese fleet in Denmark.

Frozen herring and herring fillets are much less important than frozen cod. The main exporters are those remoter from the main markets, from which it is not easily possible to send fresh herring. They include Iceland, Sweden and Norway, and to a lesser extent Denmark and the United Kingdom. The dominant part of this sector of the trade goes to Eastern Europe, and East Germany, Czechoslovakia, the U.S.S.R. and Poland all figure prominently in the list of importers, and only a minor fraction of the trade circulates within the seaboard nations of Western Europe.

The Netherlands is now the leading exporter of salt herring, and Iceland, the Faroe Islands, Norway and Britain also participate in a much shrunken export trade. While a part of this still goes to provide the markets of Eastern Europe, these are now largely supplied by the fleets of their own nations, and West Germany is actually the biggest single market. Herring preserved by the other traditional method of smoking, have poorer keeping qualities and enter less into trade, although there is some export from the Netherlands and Britain—that from the latter going mainly to English-speaking countries overseas such as Australia.

Herring are most often canned for home consumption in different countries, and most of the export comes from Norway, Britain and Holland. The biggest single market for this more valuable product is the U.S.A., although West Germany, France and other Western European countries import lesser amounts.

By far the greatest quantities of herring products now made and traded in Europe are those of oil and meal, and in 1965 over 170,000 metric tons of oil and 450,000 metric tons of meal crossed international frontiers. The outstanding suppliers here are Norway, Iceland and Denmark. There is a concentrated flow from these countries and the system of distribution branches into nearly every country. The main single market is in Britain, which bought over 85,000 metric tons of oil and over 170,000 metric tons of meal in Europe in 1965; West Germany and France are also main importers, while the trade extends in quantity into Austria, Switzerland and Czechoslovakia in the interior, and also to countries like Greece and Bulgaria in the south-east. It is noteworthy that although the countries of Eastern Europe are now among the main herring catchers on the continent, they participate little in this sector of the trade as their main aim is to increase the supplies of human food.

REFERENCES

1 F. Bramsnaes, 'Commodity Forms in Fisheries', in *Some Aspects of Fisheries Economics*, II, ed. G. M. Gerhardsen, 1964, pp. 62, 65.

2 Based on information from Mr. S. Olsen, fisheries manager, Torshavn, Faroe Islands.

3 D. J. van Dyk, Comment in the *Economics of Fisheries*, ed. R. Turvey and J. Wiseman, 1957, p. 230.

4 European Economic Community, *Rapport sur la Situation du Secteur de la Peche*, 1966, Part II, p. 60.

5 G. M. Gerhardsen, *Fiskeriene i Norge*, 1964, p. 155.

CHAPTER XI

Distribution and Consumption of Fish in Europe

In detail, the patterns of distribution in the fish trade are still very largely separate for different countries: they have been strongly conditioned by the perishability and general lack of standardization of the product, and individual nations have their own provisions to safeguard its quality. The characteristics of the distribution chains are themselves conditioned by the perishability of the product; they are geared to provide rapid links and prompt transfer from one link to the next, and there is little stocking by wholesalers, and limited credit facilities for enterprises. This specialized distribution system, however, scarcely applies to the trade in canned fish, where the big part reaches the consumer through the general food trade. Nor does it operate in the distribution of fish meal and oil, in which consumers are farmers, and food and other industries. Some of the internal national networks by which fish traffic moves may be composed of sea links in such countries as Norway, Iceland and Portugal, in which the distribution of population is dominantly coastal; but very generally the traffic moves over land.

The general pattern is naturally for the bulk of the traffic to proceed along main routes from coastal ports, although the volume of movement decreases at a greater rate than distance from the coast, as per capita levels of consumption virtually always decrease away from the sea. The main routes from coastal ports lead to main inland concentrations of population, which act as secondary nodes in the distribution systems: and per capita levels of consumption are consistently higher in such centres and along main routes than the average for a country. This basic pattern of distribution was very largely evolved when the railway was the sole medium of rapid transport, and it permitted places at two hundred miles (and more) from coastal ports to have regular supplies of fresh fish for the first time. Distribution networks still have the same basic form, but in detail have become much more complex within the last half-century with the rapid growth of the more flexible road medium.

The distribution pattern of the railway era was particularly well developed in Britain and Germany, and (to a lesser extent) in France and Belgium: this was the natural outcome of the concentration of much of the population in cities with early industrialization in countries with coasts on the Atlantic seaboard, and with immediate access to the productive North Sea. A feature was the forging of links between coastal ports and big inland cities, along which special rapid trains plied: thus Hull and Grimsby supplied inland centres such as Birmingham and London, and Hamburg and Bremerhaven German cities such as Berlin and Cologne. In Belgium, Brussels and Anvers became the main inland distribution nodes, with Ostend the outstanding port for landings; and in France a number of inland centres such as Paris, Lyons and Nancy established links with the main ports like Boulogne, Lorient and Concarneau. At the inland centres wholesale markets were established in which big consignments were split into smaller lots for distribution by rail or by horse and cart to smaller wholesalers and retailers. The long coast and big number of landing places inhibited to a noticeable degree the number of inland wholesale markets in France, and here there have frequently been extra links in the distribution chains[1] and extra costs to consumers. By contrast Italy, although also with numerous landing places spread over a long coast, developed wholesale markets in most of her inland cities, although the growth of the rail network was later than in France. To a lesser extent, comparable patterns were developed in such countries as Sweden, where supplies to inland centres were dominated by the main port of Göteborg, and in Spain, where ports on the Atlantic coast were the main suppliers to cities like Madrid and Zaragoza. In the small coastal countries of Denmark and Holland, there was less scope for the development of inland wholesale nodes, and in them distribution has been mainly organized by coastal wholesalers.

In all countries the railway era witnessed a vigorous growth in the retail trade, with the establishment of fishmongers and friers especially in cities and towns. Coastal wholesalers also could become involved in foreign trade, especially in exporting, and in France and Germany this has also been combined with importing.

Particular ports developed spheres which they served in competition with other ports. Frequently there is a considerable overlapping of these spheres, and in Germany the fact that the four main ports were all on the short stretch of North Sea coast in the extreme north-west of the country has meant that there are no real separate spheres. In countries like France and Britain, however, marketing spheres are more distinct, although all ports compete in metropolitan markets like London and Paris.

The present pattern of fish marketing and distribution in Europe is still largely a legacy of the railway era, although the modern tendency is for especially the shorter hauls to pass to road transport. In Eastern Europe and the U.S.S.R., despite little fresh fish being carried, the railway is the main medium for a volume of fish transport which is growing rapidly: here distances are generally longer, and road haulage is less developed. Inevitably links from main Russian ports like Murmansk and Kaliningrad to Moscow and other cities, from Gdynia to Warsaw and Warnemünde to Berlin are of growing importance to fish traffic.

The broad tendency now, especially in the more highly developed countries of Western Europe, is for coastal zones to be completely served by road transport, but for fish to be sent by rail to remoter inland destinations. In France in 1960, over 70,000 metric tons were moved distances of over 150 miles by public transport services—mainly by rail, and the essential framework of the French distribution system is still that of carriage by rail to inland wholesale markets and other destinations. Italy has had a distinctive, if relatively simple, system of distribution in which 'incettatori' (buyers who gather landings from numerous small ports) play a prominent role, but some 75 per cent. of the fish traffic was estimated to have gone on public service transport in 1960[2].

Here the fact that fish are legally required to pass through inland wholesale markets is a strong conditioning factor. From the main inland wholesale markets, distribution is characteristically concentrated in their immediate vicinity: in the late 1950s in Britain, for example, a survey showed that one-third of mongers and friers were located within two miles of wholesale markets, while only one-quarter were outside twenty miles. [3]

With recent developments in transport, including the improvement of highways, the building of motorways, the increase in size and power of lorries and the stream-lining of marketing organization, the road is now challenging rail more powerfully. Although in the 1950s various costal ports, such as Nieuport and Blankenberge in Belgium, stopped rail fish consignments, [4] in that decade rail transport was re-organizing with some success to repel the challenge of the road, by concentrating on bulk loads and giving concession rates to merchants using rail only; at the end of the 1950s, 75 per cent. of British fish trade went by rail. [5] The further developments of the 1960s have seriously undermined the role of rail transport in fish in western Europe. In Britain, Hull and Grimsby finally abandoned rail transport after the 1966 national rail strike, while smaller ports (like Lossiemouth and Fraserburgh in Scotland) went over completely to road transport about the same time. The result is that in Britain in 1970, the only significant transport of fish by rail was from the remotest of the main fishing ports—Aberdeen—to London, a distance of over 500 miles. The distribution system in Britain from the major ports of Aberdeen and Grimsby for 1966 has been mapped [6], by showing the areas served from the main distribution points with regular consignments from fish merchants. Prominent is the closer mesh in England south of the West Riding and South Lancashire, and the fact that within the mesh are sparsely populated areas with no regular consignments: these are especially prominent in the hill areas of Scotland. Road transport, too, now dominates the situation in the smaller coastal countries like Denmark, Holland and Portugal. Even with the most efficient transport systems, there is still a very considerable margin between the price to the fisherman and that to the consumer. In Britain,

P

it has been calculated that total distribution costs were c. 90 per cent. of the price paid by port wholesalers although for frozen fish the selling price ranged from 170 per cent. to 240 per cent. of the raw material costs.[7] Examples from Norway, where loads are often smaller and distances longer, have given distribution costs in the range of 200 per cent. to 300 per cent. of those at first sales.[8] In the neighbourhood of landing points margins are naturally considerably narrower, on account of both shorter overland hauls and more direct links with quayside markets.

In the modern pattern of trade, there has been greater centralization of marketing for first sales and for processing at central points—the counterpart of a more flexible transport system, and of the increased size of business units in the fish trade. In this, road transport from scattered landing places to the bigger ports is near-ubiquitous. It can be instanced in the cod fisheries of North Norway, where catches are gathered for freezing, in the concentration of Belgian processing at Ostend, and in the overland consignment of catches for both first sales and processing at bigger ports in Britain like Hull, Grimsby and Aberdeen.

Some of the wholesale trade does cross national frontiers, as Dutch and Belgian merchants in the post-war period have had the right to buy in each others ports under the Benelux arrangements[9]. Now the E.E.C. treaty is allowing a further widening of the international wholesale trade.

In the movement of fish away from coastal markets, the position now is considerably more complex and varied, with the activities of road hauliers added to those of the railway. Fish merchants are still much more numerous in the coastal ports, although they are tending to become fewer and the situation to be more dominated by big concerns. There is a growing tendency, especially in areas nearer coasts, for inland retailers to get supplies direct from port wholesalers: in the late 1950s in Britain, 60 per cent. of fish retailers dealt thus,[10] and it is also prevalent in France, Belgium and West Germany,[11] while in Holland all wholesalers are concentrated at the ports. In West Germany, however, with the greater distances of most of the country from the sea, inland wholesalers buy some 60 per cent. of the total, although they are considerably fewer than those at the ports.[12] Although at smaller ports retailers may buy big proportions of landings, in the overall picture this is a minor feature of the trade, and in the 1950s in Britain accounted for only about $7\frac{1}{2}$ per cent. of the total.[13] In some major ports, such as Ijmuiden in Holland, retailers are not allowed access to the market.

In addition to the consigning of fish from minor to major ports, an additional feature of the modern situation is the direct consignments from minor ports to inland markets. In Britain this has led to the growth of some new ports like Kinlochbervie and Ullapool in Scotland, which were unserved by the railway. In some cases, much of the coastal wholesaling may be in the hands of fishermen's co-operatives, as in Norway and on the Baltic coast of West Germany. Imports may also go direct to inland markets, and these include the frozen fish dispatched from Norway and Denmark to Britain and West Germany. Significant too is the development of vertically integrated organizations, which deal with the fish right through from catching to retail distribution: these include the German 'Nordsee', the Italian 'Genepesca' and the British combines of the Ross Group and Associated Fisheries.

A characteristic of the structure of retailing is that a considerable fraction of the trade goes not through specialized fish shops, but through more general food retailers, and fish are often sold in combination with items like poultry, or by greengrocers. In Britain in the late 1950s about 30 per cent. of retailers came into this category, and in Germany about 25 per cent. For such establishments, the fish trade has been especially valuable in winter, when supplies of many other foods are lower. The trend now is for the reduction in the number of retailers, and for an increase in the part going by mobile vans. In Britain in 1950 there were 13,750 friers compared with 17,000 in 1927;[14] and although a steep post-war decrease in numbers of retailers has been reported in West Germany in 1960, it was estimated that one-third had inadequate profit margins, showing that equilibrium was still far from being attained. This trend is largely compensated for by a growing proportion of fish sales being handled by the general food trade: this tends to apply especially to the higher-value items like canned varieties and frozen fillets, and has been stimulated by the growth of supermarkets. In West Germany in 1960, the general food trade was handling about 25 per cent. of the total fish trade, but 55 per cent. that of fish products (i.e. excluding whole fish). A significant and growing part of the traffic now goes from coastal wholesalers direct to the bigger catering establishments, often under contract, and in West Germany this accounted for about 10 per cent. of the total.[15] An essential part of the trends in retailing is the establishment of cold chains for the handling of frozen fillets, shell-fish and other quality products. It has been stated, however, that European cold chains are relatively poor, and that the per capita consumption of frozen food in even the leading European countries is only about 10 per cent. of that in the U.S.A.[16] It is still quite rare in Europe for fish to be branded as a guarantee of quality. There is, however, some growth of container traffic and paletization, especially in West Germany.[17]

Surveys in West Germany have given additional details on the patterns of distribution discussed here. Thus the specialized fish trade handles bigger proportions of the total in the coastal zone and in large towns, where the bigger effective market justifies more full-time specialization: in coastal areas about 95 per cent. of the fresh fish handled was by the specialized trade, while in the southern part of the country the proportion was 50 per cent.; and for the country as a whole, in towns of over 50,000 population 80 per cent. of all fish was handled by the specialized trade.[18]

PATTERNS & LEVELS OF CONSUMPTION

Patterns and levels of consumption are related to the distribution systems already discussed, and vary on a series of scales from the local to the continental. On the global scale, Europe vies with the seaboard areas of south-east Asia for the highest per capita levels of fish consumption in the world. Although Europe is much better fed than these nations, the latter have access to large areas of productive continental shelf, and have very little alternative source of animal protein. The per capita fish consumption in Europe, too, is generally considerably higher than in the highly developed countries of the U.S.A., Canada, Australia and New Zealand, in which very high proportions of animal fats and proteins come from other sources.

Within the European nations, consumption levels fall away from the open ocean, and also (though less markedly) inland from the Mediterranean and Baltic Seas. The highest national levels are found in the countries of the Atlantic seaboard which have a long tradition of fishing, and in which for reasons of terrain intensive stock farming is difficult and restricted. Estimates of per capita consumption vary, and the most dependable source is probably that of a F.A.O. study spread over the years 1960 to 1962:[19] outstanding are Norway, where the annual average was 61·3 kg./capita, and Portugal with 48·4 kg./capita. Iceland's consumption at c. 100 kg./capita appears to be the highest in the world.[20] Although even in Norway and Portugal fish account for under 3 per cent. of the calorie intake, it accounts for 12 per cent. and 18 per cent. of total protein respectively, and in Portugal provides nearly one-half the animal protein. Spain, Sweden and Denmark all have average annual consumption rates of over 25 kg./capita, and the figures for Finland and Greece are both in the neighbourhood of 20 kg./capita. In the remainder of Western Europe, in East Germany and the U.S.S.R., the levels vary between 10 kg./capita and 20 kg./capita, although in the economically advanced countries like Britain and West Germany this is a reflection of preference for alternative foodstuffs rather than inability to buy fish. In Central and Eastern Europe, consumption levels fall below 10 kg./capita, and in Bulgaria, Roumania and Hungary below 2 kg./capita.

Within the continent there are opposing trends in levels of consumption. In Western Europe, the trend is generally for fish consumption to stagnate or fall, especially in the more prosperous industrial nations, in which consumer preference tends to move towards more meat and eggs as incomes rise. Thus in Britain per capita fish consumption has actually halved since the 1920s,[21] and Germany and Belgium have also shown downward trends. Even so, fish is generally cheaper than other sources of animal protein and consumption is rising in Spain, where it has proved the most economic source for expanding the supply of animal foodstuffs, and also in France, where between 1938 and 1959 consumption rose from 9·7 to 12·8 kg./capita.[22] There have also been big proportionate increases in most of Central and south-east Europe with growing volumes of inward trade, although the absolute levels of consumption are still small. The most rapid increases have been occurring in the planned economies of Eastern Europe, where the U.S.S.R. objective of raising the average from 9·2 to 14·7 kg. in the seven years up to 1965[23] appears to have been substantially achieved. In the U.S.S.R. the price for fish is generally one quarter or one-third that of meat, and this largely explains the rising demand and consumption. Within these general trends, the consumption of higher-value products has been everywhere increasing. Thus although the total consumption has fallen in Belgium, the consumption of ground fish (as opposed to herring) rose by 40 per cent. to 5·2 kg. in the 20 years to 1957,[24] and in Portugal the per capita consumption of preserved varieties rose by 70 per cent. to 5·8 kg. in 1955.[25]

Price elasticities give some measure of the changing trends in fish demand. These tend to fall with rising income in the more highly developed countries, and in Britain figures from the National Food Survey Report in 1955 indicated a fall-off from those immediately before the Second World War. In Britain the income elasticity for fish has been estimated at 0·38 (i.e. when income rises a given amount, the fraction spent on fish rises by 38 per cent.), which is fractionally higher than for meat or food generally. For different items, elasticities vary widely: the demand elasticity for white fish is higher than for pelagic, and for fresh fish it is higher than for cured.[26] In West Germany the income elasticity for cured herring is actually around zero, although for the higher value items like shell-fish it is above 1·0. British estimates put the income elasticity for shell-fish at 1·18, processed white fish at 0·64 and for fresh herring at 0·07. In Germany also it has been shown that the income elasticity varies with the size of household, being 0·1 for those of three persons but 0·5 for those of four persons.[27]

Fish consumption fluctuates notably more than that of competing animal foodstuffs, and variations are observed both during the week and with the seasons. In Britain the peak of buying is from Monday to Thursday, but on most of the continent buying peaks occur at the end of the week, and are especially marked in the Roman Catholic countries of Southern and Central Europe, where the tradition of a Friday fish diet largely persists. Lesser fluctuations are seen in countries like Britain, but in Germany over 80 per cent. of all fresh fish is bought on Thursday and Friday.

In seasonal fluctuations, the higher fish consumption in Lent, with an Easter peak again is most prominent in the centre and south of the continent, but again also occurs in such countries as Britain. Studies in West Germany have shown that fish consumption begins to recover from its low summer period in October, and continues at high levels till Easter; the peak for fresh fish consumption is in the two months up till Easter, while preserved varieties are proportionately more important from autumn till February. With the less certain quality of fish in summer, and the greater availibility of alternative foods, the number of persons consuming no fish rises 75 per cent. in summer.[28]

Within different countries there are considerable variations in consumption levels. These vary mainly with location in relation to the coast, and to inland distribution centres; and there are corresponding variations in numbers of retail outlets.

It has been estimated that fish consumption in Norwegian fishermen's households is over 100 kg./capita, and more than twice the national average;[29] and similar indications of differences between coast and inland are known in such countries as Portugal, Sweden and Britain. The situation in West Germany is probably the clearest for which detailed information is available. Here in the "Land" of Schleswig Holstein and the cities of Hamburg and Bremen in the north-west, annual consumption in 1954 was over 30 kg./capita, but there was a progressive decrease to the south which was especially rapid in the areas behind the coast; thus Niedersachsen had an average of 18 kg. and Nordrhein-Westfalen of 11 kg. while in the extreme south consumption in Bavaria was 5 kg, and in Baden-Württemberg 4 kg.[30] In West Germany the average number of fish shops about 1960 was 0·75 per 10,000 population, but this varied from 2 and 3 in the north to 0·2 in the south[31] and the average city household consumes two-and-one-half times the average rural one. In the U.S.S.R. consumption levels also fall inland from the coastal areas, in which fish has traditionally featured more prominently in food supplies. Consumption levels on the Baltic coasts, for example, are around 24 kg. per capita annually, while in the central part of the country they fall to around 5 kg./capita.[32]

Differences in the composition of the fish section of the diet occur at national and local levels. Thus herring is prominent in Holland, related to a long-established fishery, and plaice is more important in the Danish diet than in any other country; and this is related to its abundance on the shallow sandy grounds around Denmark. In Portugal, the importance of sardine and tuna is related to locally available supplies, and the relative importance is high of whiting in France and of redfish in West Germany and the U.S.S.R. Consumption of shell-fish in France, and of frozen fish in Germany is high. At the more local level the example of Britain shows hake to be more popular in western areas, which are nearer the grounds where it is exploited, while haddock and herring are more popular in the northern part of the country and saithe, whiting and dogfish have higher sales in the south.[32] In Germany the most frequently consumed varieties in the coastal areas are plaice, herring, mackerel, sprat and Baltic cod, while inland these give way to cod, redfish, saithe and haddock.[33]

REFERENCES

1 European Economic Community, *Rapport sur la Situation du Secteur de la Pêche*, 1966, Part II, p. 150.

2 European Economic Community, *Ibid.*, p. 188.

3 R. A. Taylor, *The Economics of White Fish Distribution in Britain*, 1960, p. 107.

4 O. Vanneste and P. Hovart, *La Pêche Maritime Belge*, 1959, p. 159.

5 R. A. Taylor, *Ibid.*, p. 175.

6 P. Jüngst, *Die Grundfischversorgung Grossbritanniens*, 1968, p. 249.

7 R. A. Taylor, *Ibid.*, pp. 73, 161.

8 K. Meltvedt, 'Notes on an Attempt to analyse Costs in Fish Marketing in Norway', in *Some Aspects of Fisheries Economics*, ed. G. M. Gerhardsen, 1964, vol. II, pp. 117–18.

9 European Economic Community, *Ibid.*, p. 174.

10 R. A. Taylor, *Ibid.*, p. 64.

11 I. Bowen, 'Port Markets', in *The Economics of Fisheries*, ed. R. Turvey and J. Wiseman, F.A.O., 1957, p. 144.

12 H. Göben, 'Marketing Channels for Fish in the Federal Republic of Germany, and some changes in their Structure', in *Some Aspects of Fisheries Economics*, 1964, p. 64.

13 R. A. Taylor, *Ibid.*, p. 164.

14 R. A. Taylor, *Ibid.*, pp. 134–40.

15 H. Göben, *Ibid.*, pp. 163–69.

16 G. Lorentzen, 'The Freezer Chain—Problems and Possibilities', in *Some Aspects of Fisheries Economics*, vol. II, pp. 83–90.

17 European Economic Community, *Ibid.*, p. 194.

18 H. Göben, *Ibid.*, p. 164.

19 F.A.O., *Fisheries in the Food Economy*, Basic Study no. 18, 1968, pp. 1–6.

20 P. F. Meyer-Waarden and A. von Brandt (ed.), *Die Fischwirtschaft in der Bundesrepublik Deutschland*, 1957, p. 62.

21 E. A. R. Syms, 'Consumers Education and Sales Promotion', in *Some Aspects of Fisheries Economics*, vol. III, p. 4.

22 D. Remy, 'The Fish Industry of France', in *Atlantic Ocean Fisheries*, ed. G. Borgstrom and A. J. Heighway, 1961, p. 150.

23 G. Borgstrom, 'The Atlantic Ocean Fisheries of the U.S.S.R.', in *Atlantic Ocean Fisheries*, 1961, p. 310.

24 O. Vanneste and P. Hovart, *Ibid.*, p. 172.

25 R. J. Houk, 'The Portuguese Fishing Industry', in *Atlantic Ocean Fisheries*, 1961, p. 170.

26 R. A. Taylor, *Ibid.*, pp. 218–20.

27 H. Göben, 'Factors Affecting and Elasticities of Demand for Fish in the Federal Republic of Germany', in *Aspects of Fisheries Economics*, vol. II, p. 179.

28 H. Göben, *op. cit.*, pp. 175–77.

29 G. M. Gerhardsen, 'Notes on the Fisherman-Farmer Way of Life, especially in Norway', in *Some Aspects of Fisheries Economics*, vol. II, p. 11.

30 P. F. Meyer-Waarden and A. von Brandt (ed.), *Ibid.*, p. 63.

31 N. P. Sisoev, Personal communication.

32 H. Göben, *Ibid.*, p. 164.

33 R. A. Taylor, *Ibid.*, Appendix I.

34 H. Göben, *op. cit.*, p. 184.

CHAPTER XII

Conclusion

Even now, when their traditional character has considerably changed, the fisheries still constitute one of the most distinctive activities and employments in all countries. Although their absolute importance, as measured by quantities and values of landings, has grown at an accelerating rate in the modern period, their relative importance has steadily decreased, as manufacturing industry and service employments have come to occupy progressively larger places in national economies. Fisheries now tend to be in a particularly weak position economically, and this is largely due to there being limited effective control on the exploitation of international resources on which there are biological ceilings on yield. Linked to this are the important issues of conservation and national fisheries limits; the modern situation is essentially one of challenge, with the twin problems of the long-term maintenance of fish stocks and of the maintenance of economic viability within the industry.

CONSERVATION

There has now been a century of progress in the gathering and interpreting of data on exploited fish populations by marine scientists. A certain amount of international regulation on fisheries has also been achieved, although it becomes increasingly clear that an extended framework of regulation is required in the interests of long-term conservation.

The main results of marine scientific research, as far as they affect European fisheries, were discussed in Chapter III. Most of the work in this field has in fact been done by the European nations, and fisheries science owes much to the pioneer work of men like Frank Buckland in Britain and Einar Lea in Norway. As early as the 1860s, Buckland was proclaiming the necessity of allowing young members of stocks to escape during fishing operations, to allow them to attain maturity before being caught.[1] In the early years of the present century, Lea was able to demonstrate the wide fluctuations in annual recruitment to fish stocks—particularly that of the Norwegian winter herring.[2] During the present century there has been a large-scale build-up in fisheries science on a basis of international co-operation, and all the main fishing nations have now their own institutes pursuing research programmes. In 1902, the I.C.E.S. organization (International Council for the Exploration of the Sea) was initiated by eight founder members—Denmark, Finland, Germany, Holland, Norway, Russia, Sweden and the United Kingdom. This organization has since 1906 published a statistical bulletin on fish caught, and the organization has since been joined by France, Belgium, Italy, Spain, Portugal, Iceland, Ireland and Poland: i.e. all the European nations exploiting the fisheries of the north-east Atlantic area, with the exception of East Germany, are now members. The increased pressure brought to bear on several of the more important stocks has been a stimulus to the expansion of research.

Intensity of exploitation of the stocks of the more distant North-West Atlantic has been less. There was, however, an international scientific council for the area throughout the inter-war period, but the I.C.N.A.F. Convention (International Convention for North-West Atlantic Fisheries) was not signed until 1949. When West Germany and the Soviet Union joined this convention in the late 1950s, it came to include all the nations involved in any significant scale in the fisheries of the area: the U.S.A. and Canada are members along with the main fishing nations of Europe. Figures for the catches in the I.C.N.A.F. area have been published since 1954   The General Fisheries Council for the Mediterranean was instituted after the Second World War under the auspices of F.A.O., and has since co-ordinated scientific research in the area; a move is now afoot to collect data of catches in a standardized international form.

From the end of the last century, a main objective of fisheries scientists has been to gauge the effect of exploitation on fish populations, and this becomes especially important when over-fishing occurs. Evidence for the over-fishing of some North Sea species was officially recognized in Britain by a select committee as early as 1893.[3] Debate continued on the extent of the effects of fishing operations, but the First World War was to give conclusive proof: after a five-year period of greatly reduced fishing activity, yields leapt notably upwards With the increasing accumulation of knowledge on the population dynamics of exploited fish populations in the inter-war period, the main debate was to shift from the effects of fishing towards the formulation of measures to control those effects Now intensive exploitation has become so widespread that there are examples of species being affected throughout the I.C.E.S. area, and fears have been expressed in the I.C.N.A.F. area, especially for the cod stocks.

The earliest attempts at regulation of fisheries operations were imposed at national levels, and largely represented political solutions to disputes rather than moves towards conservation. The fixing of herring seasons by the Dutch, the stipulation of gears which could be employed in different sea areas at Lofoten, and the closing of the Moray Firth and Firth of Clyde to British trawlers were all basically of this kind. There were, however, some attempts to impose regulations on inland fisheries in the interests of conservation from the Medieval period, and the first comparable measures in sea fisheries were taken in the inter-war period by Denmark and Britain in imposing minimum legal limits on the sizes of certain white fish species which could be landed. At the London conference of European fishing nations in 1937, international recommendations were put forward on minimum mesh sizes for all types of bottom drag-nets for demersal fish; the war delayed action on these, but a North-East Atlantic Fisheries Convention was drawn up in 1946 and they were ratified by all I.C.E.S. countries by 1954. A permanent commission was set up for the enforcement of such measures. They applied to the sea areas between long. 42°W. and long. 32°E., and north of lat. 48°N., but excluded the Baltic. The limit set to mesh size was 75 mm., but 110 mm. in Arctic and Icelandic waters: this regulation recognized the fact that most species reach greater sizes in northern waters.

The Rome conference of 1955, held under the auspices of the

United Nations, showed a widening interest and dealt with principles of fisheries conservation on a global scale, and subsequently a new convention of wider scope was drawn up for the North-East Atlantic area in 1959. The area covered by the convention was extended south to lat. 36°N., and east to long. 51°E. The institution of three separate committees to deal with conservation in the Arctic, Faroe, and North Sea/West coast of Britain waters indicates an effort to adapt measures on a regional basis, and a fourth committee was later formed to deal with waters between lat. 48°N. and lat. 36°N. The 1959 convention also imposes on the commission the duty of keeping fisheries under review; it enables it to make recommendations on closed seasons and closed areas, and on measures to regulate fishing effort and size of catch. One of the most important effects of these wider provisions was to give powers to initiate measures to conserve pelagic stocks as well as demersal. Recommendations were made in 1969 for amendments to mesh size regulations, both to alter mesh sizes and to cater more for the regional variations in fish stocks. A refinement is the allowing of smaller meshes for nets of modern synthetic fibres, through which fish slip more easily; and recommended mesh sizes vary from as little as 60 mm. for synthetic fibre trawls in the southern area to 130 mm. for trawls of manilla fibre in the Arctic. Although the latest proposals for mesh sizes for the catching of demersal fish represent provisions to cater for regional variations, even these are compromises in that the nets are used in catching a series of species for each of which there is a different optimum size. An agreement of 1970 reached by the North-East Atlantic Fisheries Commission for the institution of closed seasons for herring represents the first step in conservation for this important species.

The 1959 convention also includes measures for the encouragement of fish husbandry in the seas. To date fish farming has been successful with species like trout in inland waters, and with mussels and oysters in estuarine situations. Fish husbandry in the open seas, however, is a much more difficult and complex problem; although some species have already been reared in big numbers through the critical infant stages, it is at present difficult to envisage much augmentation of commercial stocks by this means.

An issue which has come increasingly to bear on the conservation of the world's organic resources is that of environmental pollution. This could be especially serious for fisheries in the long term, as the oceans are generally regarded as the most suitable dumping grounds for toxic materials and for other harmful refuse. While the immense volume of ocean waters ensures a great dilution of harmful waste, there are now indications that it is possible for these to build up to levels which can be dangerous in at least some regions: and the effect of toxic substances like mercury which can build up in the tissues of fish and not be secreted is especially harmful. In the short term, the greater dangers are in inland waters where the dilution of waste is much less; this already has caused concern in Europe, and has led to the suspension of some fishing activity. In the seas, the greatest dangers are in the enclosed seas of the Caspian, Mediterranean and Baltic, and toxic substances have been found in measurable concentration in some fish stocks in all of these. On the west side of the Atlantic, there have already been cases where fisheries off eastern Canada and the eastern U.S.A. have been suspended because of toxic substances being found in commercial fish stocks, although this in part reflects the very wide safety margins that are considered desirable. With the scale and character of modern industrial development on many European coasts and rivers, it is obvious that there are dangers to the fisheries on the continental shelves of the north-eastern Atlantic, and the area of the southern North Sea is in particular danger. It is evident that the situation requires systematic international monitoring and regulation to ensure safety.

NATIONAL FISHERIES LIMITS & THE INTERNATIONAL LAW OF THE SEA

An important issue for fisheries has been the international law of the sea, which has been in Europe a matter of controversy for centuries. Prior to the sixteenth century, the basic concept was that of *mare clausum*; by this various countries claimed exclusive rights over particular sea areas, although the purpose of trade was generally more important than fisheries in those claims. The doctrine of international freedom of the seas (*mare librum*) is associated with the name of the the Dutch lawyer, Hugo

Grotius, who published a book with this title in 1609;[4] although the concept was slow to gain acceptance, it eventually became internationally recognized as being in the interests of all nations in a period of expanding seaborne trade. At this period, however, fisheries interests were important in northern waters, and fishing limits were claimed around Iceland and on the Norwegian coasts; for a period the former extended to 24 miles.[5] The general acceptance of the principle of international freedom of the seas was not fully achieved until the nineteenth century. By general agreement, territorial waters in both Europe and the North-West Atlantic were restricted to the so-called 'cannon-shot' distance of three miles; and the only limited exceptions to this principle were the states of France, Spain and Russia.

In the nineteenth century, the limits to the territorial sea were also generally recognized as being fishing limits, and the usual base line from which they were measured was the low water mark of spring tides. Increasing pressure on fisheries resources in the present century, however, has stimulated some re-assessment of fishing limits, and the situation since 1945 has been further complicated by claims to mineral rights (especially for oil and natural gas) on the continental shelf. At a conference held under the auspices of the League of Nations in 1930, the Netherlands, Poland and France all claimed fishing limits of six miles, while Norway, Sweden, Finland and Iceland claimed that their limits had never been reduced below four miles.[6] Since then it is significant that the main proponents of extension of fishing limits have been the lands in high latitudes with rich near-shore grounds (Iceland, Norway and the Faroe Islands); and the leading objectors have been countries with big distant-water fisheries interests in the Arctic (mainly Britain and Germany). It is noteworthy that within Britain there have been divided interests on this matter: in Scotland most of the fisheries are inshore, and most interests have favoured protection of nearer waters, while English fisheries are dominated by the distant-water sector, whose interests were best served by fishing limits at minimal distances from land.

In post-war times, not only the extent of fisheries limits but the methods of delimiting the base-lines have also been major issues. Both Norway and Iceland have had the right to use point-to-point base-lines, linking peninsulas and off-shore islands, upheld by the International Court of Justice at the Hague, when Britain contested their claims. Britain has been a main proponent of 'historic rights' in fisheries, whereby established practice with regard to the use of fishing grounds would have legal recognition; but historic rights now are being recognized only over a transition period in the extension of fishing limits.[7] The International Law Commission, which had been working over the post-war period, brought out its final recommendations on fisheries in 1956. While stating that the final goal should be universally agreed laws, it recognized the rights of coastal states with special interests to impose unilateral regulation on fisheries in the off-shore zone. Iceland in 1958 extended her limits to twelve miles to help safeguard vital economic interests, and since then the extension to this distance, measured from point-to-point base-lines has become general, and was formalized for the North Sea area by a treaty of 1964.

The extension of limits to ensure the interests of particular countries has now been partly overtaken by other developments in the shape of the forging of international economic blocks. The E.E.C. finally agreed on its fisheries policy in November 1970, and this allowed the exploitation of the zone within twelve miles of the coasts of member states by all their fishermen on equal terms. With the prospective entry into the E.E.C. of Britain, Norway, Denmark and Ireland, it has now been agreed that the fisheries policy of the group will be reviewed. These four countries with their long coasts stand to lose considerably from the opening of their inshore grounds to fishermen of other member countries, and Norway in particular has stood out for the retention of its twelve-mile limit.

THE IMPORTANCE OF FISHERIES IN DIFFERENT COUNTRIES

Fisheries naturally vary greatly in importance in Europe, and the general tendency is for them to become less important both from north to south and from west to east. As well as differences at national level, there are also important variations at regional level, although measures of this are few, as most statistics are kept on national bases.

At international level, the best single index of comparison is that of the proportional value of fish in the gross national product (G.N.P.). Even at the highest, this proportion is relatively small, because of the dominating importance of activities in the secondary and tertiary sectors. In Iceland estimates of the G.N.P. are somewhat tentative, but recent figures show the proportional value of fish landed as around 25 per cent. of the G.N.P. In the Faroe Islands, where there is effectively value added by processing at sea (in wet salting of cod before landing), the proportion has been over 30 per cent. in the 1960s. In Norway, the rise of other sectors in the economy has depressed the value of fish to c. 2.5 per cent. of the G.N.P. In both Spain and Portugal it is between 1 per cent. and 2 per cent., and in all other countries it is now below 1 per cent.: in Denmark, for example, it is c. 0·80 per cent., in Italy 0·30 per cent., and in Britain 0·17 per cent. of the G.N.P.

If the importance of fisheries in all national economies is now minor, fish and fish products can have a more important role in national trade balances. In both Iceland and the Faroe Islands, over 90 per cent. of exports by value are fish, although in Norway they now have declined to c. 13 per cent., and in Portugal to c. 8 per cent. Developments in Denmark in the post-war period have actually led to an increase in the importance of fish in foreign commerce, although exports are at the level of c. 5 per cent. of the national total. Although Britain and West Germany are two of the leading importers in the world of fish by value (next only to the U.S.A.), the proportional value of fish in their imports in well under 1 per cent., and in no European country is fish of substantial importance in the import trade.

Another index of the importance of fisheries is the proportion of the employment they provide in different countries and regions. In Iceland in 1960, 10·8 per cent. of the population,[8] or c. 30 per cent. of the labour force were engaged in fisheries. In the Faroe Islands, the proportions are higher: a total of c. 6,000 fishermen represents 15 per cent. of the population, or c. 45 per cent. of the labour force. In Norway fishermen now represent only about 4 per cent. of the labour force, although in the 1930s they were over 10 per cent.; and Denmark and Portugal are the only two other countries in the continent in which fishermen constitute over 1 per cent. of the labour force.

Q

If data on gross product and exports are lacking on a regional basis, it is nevertheless possible to indicate something of the regional importance of fisheries in employment. Thus in the three counties of North Norway, fishermen constitute nearly 20 per cent. of the labour force. In Scotland, fishermen account for over 10 per cent of the labour force in the Shetland Isles, and 4 per cent. of that of the north-east; and in the latter area total employment in fisheries (i.e. fishermen together with workers in fish trades on land) amount to 12 per cent. of the regional labour force. At a more detailed level, fisheries are still capable of dominating local economies—not only in fishing towns and villages on all the coasts of Europe, but also in the major modern port of Murmansk, where it has been stated that the industry employs half the labour force in a population of 250,000, and accounts of three-quarters of its gross industrial product.[9]

Fisheries provide an intriguing and outstanding example of an economic activity pursued with a wide range of modern technological equipment, and yet still basically at a level of hunting rather than of husbandry in the utilization of a biological resource. The application of technology in the U.S.S.R. now includes the large-scale use of computer systems to record and organize the operations of the distant-water fleets based in the Baltic and Murmansk. Europe is the continent in which most of the techniques and practices employed in modern fisheries were pioneered, and it now most urgently faces the resultant problems of conservation of the resources. This involves not only the task of maintaining the fish stocks in the long term, but also that of rendering the industry that exploits them economically viable; and (in contrast to the situation on land) this must be done for the foreseeable future without increasing the yields that can be drawn per unit area. Modern evidence and theory would suggest that further regulation of both the composition of catches and of fisheries effort is required. The task now for international statesmen is the harmonizing of a whole series of biological and economic variables—no simple matter in an industry dominated by large numbers of *individual* vessels, generally working within a framework of both national and international competition.

Progress has been made towards a supra-national framework

of organization in fisheries, and it may be that the contemporary moves towards integration in Europe may allow something of the co-operation in the economic organization of fisheries that has been achieved in international research in marine biology. As traditional small-scale fisheries continue to decline, more elements of uniformity are appearing in different fisheries, and this should enhance the possibility of international regulation. There is, however, the accompanying danger of subjecting individual units and men to the control of a more remote central organization. The formidable task for the future is the setting up of bodies large enough in scope to organize the fisheries of Europe—or indeed the world—but sufficiently flexible to allow for regional and local characteristics. The fisheries now face the challenge of a world shrunk by modern technology and communications.

REFERENCES

1 B. Engholm, 'Fishery Conservation in the Atlantic Ocean', in *Atlantic Ocean Fisheries*, ed. G. Borgstrom and A. J. Heighway, 1961, p. 40.

2 A. Hardy, *The Open Sea*, vol. II., 1959, pp. 58, 59.

3 B. Engholm, *Ibid.*, p. 41.

4 G. Winder, 'International Territorial Limits', in *Atlantic Ocean Fisheries*, 1961, p. 50.

5 L. M. Alexander, *Offshore Geography of Northwestern Europe*, 1963, p. 10.

6 G. Winder, *Ibid.*, p. 51.

7 L. M. Alexander, *Ibid.*, pp. 101–6

8 O. Bjornsson, 'The Icelandic Fisheries', in *Atlantic Ocean Fisheries*, p. 266.

9 R. A. Helin, 'Soviet Fishing in the Barents Sea and the North Atlantic', *Geog. Review*, 1964, pp. 400–1.

Additional Bibliography

F. Bartz, *Die Grossen Fischereiräume der Welt*, Band II. *Asien mit Einschluss der Sowjetunion*, 1965.

W. Blanke, *Die Seefischerei Nordwest-Europas: Struktur und Probleme*, 1956.

G. Borgstrom (ed.), *Fish as Food* (2 vols.), 1961 and 1962.

A. von Brandt, *Fish Catching Methods of the World*, 1964.

G. H. O. Burgess, *Developments in Handling and Processing Fish*, 1965.

E. E. D. Day, 'The British Sea Fishing Industry', *Geography*, 54, 1969, pp. 164–80.

I. Hela and T. Laevastu, *Fisheries Hydrography*, 1961.

H.M.S.O., *Fish Handling and Processing*, 1965.

H. Lübbert, E. Ehrenbaum and A. Willer, *Handbuch der Seefischerei Nordeuropas* (11 vols.), 1938–55.

O.E.C.D., *Fish Marketing in the O.E.C.D. Countries*, 1952.

O.E.C.D., *Fish Marketing and Preservation*, 1965.

O.E.C.D., *Fishery Policies and Economies*, 1957–66, 1970.

O.E.C.D., *Review of Fisheries in O.E.C.D. Member Countries*, 1968.

O.E.C.D., *Fishery Policies in Western Europe and North America*, 1960.

B. Parrish and A. Saville, 'The Biology of the North-East Atlantic Herring Populations', *Oc. & Mar. Biol. Ann. Review*, 3, 1965, pp. 323–73.

G. A. Rounsefell and W. H. Everhart, *Fishery Science, its Methods and Applications*, 1953.

E. S. Russell, *The Overfishing Problem*, 1942.

A. Scott, 'Bibliography of Fisheries Economics', in *F.A.O. Fisheries Report no. 5*, 1962.

J. Traung, 'Fishing Boats of the World,' *F.A.O.*, 3 vols., 1955, 1960, 1967.

Appendix

LANDINGS, IMPORTS AND EXPORTS OF FISH,
BY QUANTITY AND VALUE, FOR EUROPEAN COUNTRIES IN 1968

COUNTRY	LANDINGS		IMPORTS		EXPORTS	
	Quantity '000 metric tons	Value '000,000 $ U.S.	Quantity '000 metric tons	Value '000,000 $ U.S.	Quantity '000 metric tons	Value '000,000 $ U.S.
Austria	4·0	3·8	75·1	21·5	0·5	0·5
Belgium	68·4	16·8	217·5*	65·8*	30·5*	14·3*
Bulgaria	56·3	n.a.	17·9	n.a.	11·9	n.a.
Czechoslovakia	13·6	n.a.	138·0†	n.a.	0·9‡	0·4‡
Denmark	1,466·8	100·5	191·9	29·9	497·3	136·1
Faroe Islands	166·3	n.a.	1·9	0·4	75·2	19·2
Finland	92·7	19·1	49·4	12·9	0·2	0·1
France	793·6	239·3	358·5	168·3	44·9	25·7
Germany (E)	295·0	n.a.	139·7‡	n.a.	n.a.	n.a.
Germany (W)	682·3	91·8	985·5	186·8	141·9	51·2
Greece	85·1‡	35·2	43·2	14·7	2·6	1·8
Hungary	29·9	43·3	69·5	n.a.	2·8	1·5
Iceland	600·7	n.a.	0·1	0·4	286·8	69·7
Ireland	53·1	7·9	24·6	6·2	25·4	n.a.
Italy	363·4	183·7	306·0	125·6	11·9	6·0
Malta	1·6	0·8	n.a.	0·9	—	—
Netherlands	323·3	62·6	425·8	71·5	198·6	83·7
Norway	2,804·1	146·8	73·1	10·6	818·7	216·1
Poland	406·7	n.a.	121·4	17·1	48·4	15·6
Portugal	559·8‡	74·2‡	77·8‡	33·1‡	98·1‡	51·9‡
Roumania	40·5	n.a.	n.a.	n.a.	n.a.	n.a.
Spain	1,503·1	313·8	171·0	37·3	96·9	49·6
Sweden	315·3	42·2	185·7	69·1	191·6	24·3
Switzerland	3·5	2·8	83·5	33·2	0·5	0·6
United Kingdom	1,040·3	64·3	965·7	277·1	75·8	29·4
Yugoslavia	44·9	n.a.	57·2	9·8	13·5	7·8
U.S.S.R.	6,082·1	n.a.	14·7	14·6	320·1	83·2

* Including Luxemburg.
† Figures for 1966.
‡ Figures for 1967.
n.a.—not available.

Index